Wind Power

Wind Power looks at the nations, companies and people fighting for control of one of the world's fastest-growing new industries and at how we can harness one of the planet's most powerful energy resources – wind power. The book also examines the challenges the sector faces as it competes for influence and investment with the fossil-fuel industry across the globe.

The wind-power business has grown from a niche sector within the energy industry to a global industry attracting substantial investment in recent years. In Europe, wind has become the biggest source of new power generation capacity, while wind is successfully competing with the gas, coal and nuclear sectors in China and the US.

The specialist wind-turbine companies that pioneered the business have gone global over the last decade, while big industrial conglomerates have entered the fray. European companies are struggling to maintain their technological and market lead in the sector, in a three-cornered battle with China and the US.

Meanwhile, the industry is fighting to drive down costs in the face of a fossil-fuel-generation industry bolstered by the onset of cheap shale gas. And wind companies continue to wrestle with a stop–go cycle of investment across the world, with some traditional markets stuttering under the impact of the financial crisis and fiscal austerity, while new power-hungry markets in Asia and South America emerge.

Wind Power analyses the industry climbers, the investment trends and the technological advancements that will define the future of wind energy.

Ben Backwell has spent most of his career covering international energy markets and finance. He worked for international news agencies and as an analyst in Houston, Caracas, New York, Rio de Janeiro and Buenos Aires, and covered the oil producers' organisation OPEC, before moving back to his native UK in 2006. He then worked as an analyst covering major oil companies before joining the renewable energy news service, *Recharge*, ahead of its launch in January 2009. He became Editor-in-Chief of *Recharge* in 2012. He has a Master's Degree in Politics from the University of London.

Wind Power

The struggle for control of a new global industry

Ben Backwell

Routledge
Taylor & Francis Group

LONDON AND NEW YORK

First published 2015
by Routledge
2 Park Square, Milton Park, Abingdon, Oxon OX14 4RN

and by Routledge
711 Third Avenue, New York, NY 10017

Routledge is an imprint of the Taylor & Francis Group, an informa business

© 2015 B. Backwell

The right of Ben Backwell to be identified as author of this work has been asserted by him in accordance with sections 77 and 78 of the Copyright, Designs and Patents Act 1988.

British Library Cataloguing in Publication Data
A catalogue record for this book is available from the British Library

Library of Congress Cataloguing in Publication data
Backwell, Ben.
    Wind power : the struggle for control of a new global industry / Ben Backwell.
        pages   cm
    1. Wind power industry.   2. Renewable energy resources.   3. Energy development.   I. Title.
    HD9502.5.W552B33 2014
    333.9'2–dc23
    2014007677

ISBN: 978-0-415-72961-1 (hbk)
ISBN: 978-1-138-84585-5 (pbk)
ISBN: 978-1-315-77476-3 (ebk)

Typeset in Sabon
by Out of House Publishing

Contents

List of figures and tables

Figures

Tables

Foreword

Wind power is the first of the so-called 'new renewables' to establish itself as a mainstream power source. Utility-scale wind projects operate in more than 80 countries, with 24 countries having more than 1,000 MW installed, supplying nearly 3 per cent of the global electricity supply. Total installations at the end of this year will be well over 300 GW, and recent IEA projections, for instance, show that wind could be supplying 15–18 per cent of global electricity by 2030.

Since a small company built the world's first commercial wind farm at Crotched Mountain near my childhood home in southern New Hampshire in late 1980, the industry has transformed itself beyond all recognition. It took 18 years to get to the level of 10,000 MW in 1998, and another 10 years to reach 100 GW in 2008. Passing 200 GW in early 2011, and now 300 GW by the end of 2013, the numbers and the contribution to the global electricity supply are mounting rapidly. From a few small entrepreneurs in the US and Denmark, wind energy has emerged as a technology embraced by most of the world's major energy companies and utilities. Wind power supplies increasingly cost-competitive, carbon-free electricity in an ever-expanding number of markets around the world, enhancing energy security, stabilising electricity prices and creating good-quality jobs.

Ben's book is a valiant attempt to nail down a snapshot of a dynamic industry reaching maturity and going global while at the same time being buffeted by myriad forces beyond its control. While no 'insider' will agree 100 per cent with every bit of his analysis, most of it rings pretty true to me. He explores all the main themes while pointing out that what the future will bring is still very much in play, and importantly, draws the fundamental link between climate policy and the development of the industry, which is not often apparent in the daily to and fro of the energy policy debate. In fact, it is this disconnect between the climate and energy policy debate both nationally and internationally that is responsible for the start–stop nature of support for wind in many key markets.

At the same time, of course, wind has made tremendous strides towards establishing itself as a mainstream power source on its own merits as a supplier of affordable, clean and indigenous energy; even in the face of massive fossil and nuclear subsidies and the lack of an effective carbon price. He makes the case very clearly that wind's greatest enemy is not its variable production, but the volatility of policy swings and roundabouts that stand in the way of delivering the maximum quantity of carbon-free electricity at the lowest possible price.

I strongly recommend Ben's book to those who want to get a broad picture of just how far wind energy has come in the past three decades, as well as the promise and challenges of the road that lies ahead.

Steve Sawyer, General Secretary, Global Wind Energy Council (GWEC)

Acknowledgements

Taking part in the growth of the wind industry, after years spent primarily covering fossil-fuel energy and finance, has been an amazing experience. As befits an industry that is still relatively young and – more importantly – one whose participants actually feel they are doing something worthwhile, I have been impressed since day one by the amount of openness in the wind industry. The idealism and pioneering spirit that is in the DNA of wind can still be felt even in the biggest corporate players in the sector.

Over the last five years, I have interviewed and hung out with some remarkable people, from some of the founders and early pioneers to long-term advocates, entrepreneurs and the tenacious engineers who day by day are out there finding new solutions to make the technology bigger, smarter and more efficient.

I have been overwhelmed by the positive response I have had when I told people I was writing this book, with a whole number of senior industry figures agreeing to be interviewed, suggesting themes, and reading and checking sections and chapters. Any mistakes of course, are entirely my own.

I would like to thank in particular Henrik Stiesdal, the CTO of Siemens Wind, one of the inventors of the modern wind-turbine industry and one of its most critical thinkers, whose feedback gave me the confidence that I was on the right track.

GWEC General Secretary, Steve Sawyer, the industry's most tireless advocate, has provided constant encouragement and inspiration since I started working at *Recharge*.

Others who contributed to this book with advice and valuable insight are consultants Christian Kjaer – who was the CEO of the European Wind Energy Association (EWEA) at a crucial period in the development of the wind industry, Robert Clover, Feng Zhao; and GWEC Chairman, Klaus Rave.

Alstom's Alfonso Faubel and The Switch's Jukka Pukka Makinen gave invaluable insight into current thinking on manufacturing systems and supply-chain management; Blade Dynamics' Pepe Carnevale and Theo Botha gave me a completely new perspective on composites and rotor blades; while much of what I have learnt about offshore wind is thanks to Areva's Julian Brown.

I want to also thank Vestas' Morten Albaek, who has refused to be satisfied with the routine and ordinary, and tried to creatively disrupt the wider energy business. As well as creating initiatives like WindMade and the Global Corporate Renewable Energy Index, Morten also found the time to help me rethink my own business plan when I was redesigning *Recharge*'s strategy and products in late 2012.

The creative background to this book has been a whole series of discussions under the auspices of *Recharge*'s Thought Leaders Club, which has brought together many of the leading thinkers in the wind sector and culminated in the first Holmenkollen Summit in January 2014. Among the many people I was able to discuss the industry with in 2013 I would like to mention Andrew Garrad – another of the industry's pioneers and one of its most compelling champions; DNV's Henrik Madsen, Bjorn Haugland and Johan Sandberg; Goldwind's Wu Gang; E.ON's Michael Lewis; Natural Power's Jeremy Sainsbury OBE; Enel Green Power's Francesco Starace; ABEEólica's Elbia Melo; Mainstream's Adam Bruce; GE's Anne McEntee; the Crown Estate's Rob Hastings; Repsol's Ronnie Bonnar; EWEA CEO Thomas Becker; and Offshore Renewable Energy Catapult CEO Andrew Jamieson.

Special thanks go also to Oliver Loenker, Michael Zarin, Luis Adao da Fonseca, Eduard Sala de Vedruna, Frederick Hendrik, Enrique de las Morenas, Anders Soe Jensen, Aris Karcanias, Per Krogsgaard, Nick Medic, Julian Scola, Vicente Trullench, Jennifer Webber, Gordon Edge, Ramesh Kymal, José Antonio Miranda, Heikki Willstedt, Malgosia Bartosik, Jayasura Francis, Isabelle Prosser, Ricardo Ferreira, Vineeth Vijayaraghavan, Jonathan Collings, Ekow Monney and Bruce Douglas.

Closer to home, NHST Media Group CEO, Gunnar Bjørkavåg, has been an enthusiastic supporter of *Recharge* and renewables. *Recharge*'s Commercial Director, Angelo Ianelli, has been with me through thick and thin and on a series of highly amusing journeys around the world, as has Technology Editor, Darius Snieckus. Thanks also to Sidsel Norvik, Chris Hopson, Karl Erik Stromsta, Andrew Lee, Richard Kessler, Brian Publicover, Milton Leal, Dominique Patton, Leigh Collins, Simon Bogle, Julian Stuart and Anamaria Deduleasa.

Finally, thanks to my wife Melissa, and my two sons Dimitri and Thierry.

Introduction

The struggle for the global wind-power market

From the arid planes of Gansu province in China to Oklahoma, to the rough waters of the North Sea, wind power is on the march.

Once a small industry selling mainly to farmers and cooperatives, and dominated by a couple of small, mainly Danish firms, the sector has become a key investment area for industrial heavyweights such as Siemens, General Electric, Mitsubishi and Areva, in the search for zero-emission power generation.

In key markets in Europe, the US and China, wind has gone from being a niche power source, to one that is increasingly challenging 'conventional' energy (coal- and gas-fired thermal generation and nuclear), and along the way disrupting the traditional business model of the world's big power utilities. In 2012, for example, wind was the biggest source of new power generation in the US – with a record 13.2 GW of new installations – for the first time ever, and it had already won this place within the EU back in 2008.

Key to the industry's rise have been government incentives and targets around the global attempt to limit carbon emissions, including landmark renewable-energy targets in the EU, national feed-in tariff and green certificate schemes, state portfolio standards and the production tax credit (PTC) in the US. In some major European countries, large-scale wind power – and in particular the growing offshore sector – is practically the only way to reach their 2020 targets and close the gap in capacity caused by the decommissioning of older coal, gas and nuclear plants; while for the US and China, wind is essential if the two countries want to make any progress on stemming their rise in carbon emissions. But government support has not been the whole story, and is likely to become a less important part of the story going forward.

The impact of the March 2011 accident at the Fukushima Daiichi nuclear plant in Japan and the subsequent decision by Germany and other nations to shut down or put on hold their nuclear industries, strongly added to wind's momentum. In many countries, a lack of available fossil fuels – and the high cost of importing them – makes wind a logical way to add domestically produced power quickly, cheaply and cleanly, while creating local jobs. Politicians in India are fast waking up to the fact that cost and infrastructure bottlenecks make building new coal plants a pointless exercise; while in China, growing public pressure over air pollution means that policymakers are looking more and more favourably on wind.

Long seen by traditional voices within government and the power industry as too costly, it has been shown in places like Brazil's northeast and Texas in the US that wind power can compete on price with conventional power.

In Brazil, record capacity factors have allowed wind-power projects to win power contracts at prices as low as R$87.77/MWh (or around US$42/MWh[1]), while gas producers were unwilling to take part in a government power tender with a floor price of R$140/MWh (US$67/MWh).

In the US, the wind industry has shown itself remarkably resistant to lower gas prices, despite all the fanfare over the 'shale gas revolution'. The sector has seen record growth in the face of Henry Hub natural gas prices at lows of US$2–3/mn BTU.

Even in Australia – the world's largest exporter of coal – wind energy (AUD 80/MWh (US$75/MWh)) is beating new coal (AUD 143/MWh (US$134/MWh)) and gas-fired (AUD 116/MWh (US$109/MWh)) power plants. Even excluding the carbon cost, wind energy cost 14 per cent less than coal and 18 per cent less than gas.

Moving into the mainstream brings new challenges, however. As Angela Merkel told Germany's Bundestag just before her re-election as Chancellor in September 2013, renewables no longer occupy a niche, 'but are part of the overall generation mix', adding 'that leads to entirely new problems'.

As wind's weight in the generation mix has grown, so has resistance within the big utilities, as well as more 'conservative' (with a small 'c' of course) personnel within government ministries and grid operators.

Utilities like Iberdrola, RWE and E.ON have been key investors in wind energy, but they have found that the growth of the business has increasingly affected their traditional businesses, pushing mostly gas-fired power generation off the grid when the wind is blowing, for instance. Company executives have found themselves pushing for new government support for their fossil businesses, calling for limits to be put on wind and solar deployment, or even calling for new taxes on renewables. Meanwhile, some of the largest players in the power sector have decided to abandon nuclear power. In September 2011, Siemens' CEO Peter Löscher said that the company would 'no longer be involved in managing the building or financing of nuclear plants'. The same message came from CEO Peter Terium of Germany's second-largest energy producer, RWE: 'The nuclear power chapter has come to an end for us', he announced in June 2012.

Power system operators and government officials have raised increasing opposition to wind, due to the effects of large-scale amounts of wind power on system balancing. Rather than pursue the kind of long-term solutions which are necessary for the full modernisation of power systems – cross-border power markets, a fully interconnected international grid (the 'supergrid'), storage technologies and smart metering – officials use the term 'intermittency' as a bogeyman to slow down change.

Added to this has been a new focus on costs, fuelled by the long enduring Eurozone crisis of the early part of the decade. In some countries, like Spain and Portugal, fast deployment of renewables has turned into a series of attacks on the sector, including retroactive changes to operators' contracts, as governments have tried to deal with new bankruptcy and deficit-generating tariff systems. In others, like Germany, government actions have been more considered, but there is an ongoing debate about who bears the brunt of increased power-system costs.

There is a parallel debate within the wind industry itself, about whether the sector should continue to call for government support, for instance the continual renewal of the US PTC, or whether it should aim for some kind of mid-term phase-out of subsidies. The debate is complex, however, as traditional fossil-fuel power sources enjoy a myriad of different types of open and hidden payments, something to the tune of US$500bn[2] in 2011, according to IEA Chief Economist Fatih Birol, who calls fossil-fuel subsidies 'public enemy number 1' in terms of the world meeting its CO_2 targets. Wind industry officials argue that it would make no sense to give up any support the sector receives in the short term as long as the competition is receiving its own subsidies and does not pay for its carbon emissions.

Meanwhile, public opinion has remained resolutely positive about wind, despite politicians sensing that renewables can be an easy scapegoat to blame for unsustainably expensive power systems, and the activities of a section of the conservative press which border on the pathological. A series of hysterical campaigns, which claim everything from that turbines are inefficient to that they blight the landscape or even make you ill, have spectacularly failed to change the public's gut feeling that power generated from an entirely natural and free source is a good idea.

We have come a long way since the first groups of inventors and hobbyists who started the modern wind industry. The first chapter in this book looks at the creation of the world's wind market and the rise of the mainly European wind companies – led by Denmark's Vestas – to global status. Chapter 2 looks at the entry of big global engineering giants – like Germany's Siemens and the US's GE – into the business as it gained momentum, and the entry into the business of the big power utilities, which according to who you speak to, have been key promoters of and key barriers to the growth of the industry. It also looks at the take-off of the sector in the US, which – although hardly a hawk in terms of climate negotiations – has consistently been the world's single largest market for renewable energy – until the rise of China, which we tackle in Chapter 3. In industrial terms, the battle between European, US and Chinese players is one of the key themes of this book.

Following China, in Chapter 4 we look at some of the big new players, namely Brazil, where the industry is seeing record capacity factors, rock-bottom prices and breakneck growth; and India, whose current power mix and growing economy gives it enormous potential. Arguably, it will be the success or failure of wind in defining the energy economies of these countries that will decide whether the struggle to limit carbon emissions can be successful.

We then look at the fast-growing offshore sector in Chapter 5. Over the last decade, offshore wind has become one of the fastest growing areas within the wind industry. From small beginnings in a couple of pilot programmes in Denmark and the UK, offshore wind now constitutes one of the greatest engineering challenges of our time, as projects grow to a massive scale – the largest of the UK's 'Round III' areas has a potential of 9.2GW – turbines grow to skyscraper size and wind farms move out into water depths of up to 70m. Offshore wind costs are still significantly higher than for onshore wind; and yet there are strategic factors pushing forward the growth of the sector in places like the UK, where planning and space constraints on land combined with legally binding emission reduction targets and the exorbitant costs of building nuclear power create what is virtually an imperative to build offshore. Reliability improvements, scale and growing expertise in the installation process will push offshore's costs steadily downwards, the industry argues. And while the vast bulk of offshore wind is currently in the waters of the UK, Germany and Denmark, other countries, notably China and Japan, have the capacity and the incentive to grow their offshore sectors aggressively, something which can already be attested by the appearance of a series of new Chinese mega turbines, and Japan's Mitsubishi's formation of an offshore joint venture with Vestas.

In Chapter 6, we see how the precipitous nature of wind's growth into a major global manufacturing left it ill prepared for the aftermath of the disastrous Copenhagen climate summit in December 2009. The failure of the talks on the home ground of the world's most hawkish countries in terms of taking action against emissions was a profound shock to many who took part, including the writer of this book.

Taking their signal from the top, sceptical politicians and incumbent energy lobbies stepped up criticism of the climate agenda and renewables. Wind was no longer the sexy industry on the front cover of *Forbes* magazine. The failure of talks on a binding climate agreement was not the only problem, of course. The Copenhagen fiasco also coincided with the effects of the subprime crisis being felt in earnest. As a capital-intensive sector, the liquidity squeeze hit wind energy hard compared to other sectors. Commercial financing for wind farms in Western markets largely dried up, while the utilities that had been largely driving expansion began to feel their balance sheets tighten. And, as the subprime crisis began to morph into the Eurozone crisis, key areas for wind expansion such as Spain and Portugal slammed on the brakes, while others began to slow down. Coupled with uncertainty over government support, the arrival of cheap shale gas in the US, and the onset of massive grid curtailment issues in China, the wind industry was facing a perfect storm.

Figure I.1 Ben Backwell chairs a panel of major wind-turbine developers and manufacturers in Beijing, October 2013 (Jeffrey Lau, *Recharge*)

Not that the industry stopped growing of course. But the slowdown in expansion and the continued entry of new players and manufacturing capacity into the industry meant that by 2011, some of the sector's most important players were facing huge overcapacity, bloated workforces, big debts, falling turbine prices and rock-bottom equity values. How some of the key players in the industry have coped with the post-Copenhagen hangover is the subject of Chapter 7.

In the final chapter of this book, I look at some of the key challenges facing the industry. The key among these is the struggle to reduce the costs of wind power and push fossil fuels out of the market by reaching what is rather simplistically known as 'grid parity'. Grid parity is a loaded question, as it is intimately bound up with wind's struggle to maintain political support and create a true level playing field in the global energy industry, where fossil fuels and nuclear energy still receive massive subsidies.

A second challenge is consolidation within the industry. Industry executives and analysts have been talking about consolidation for years, and there is broad agreement that some level of consolidation is necessary to raise the level of mechanisation in production and create more efficiencies of scale, thus allowing costs to go down. However, there have been so far only halting signs that widespread consolidation is taking place – at least in terms of the number of major turbine manufacturers. Whether this process is about to speed up, what should be the future model for wind turbine manufacturing and which companies are equipped to survive are the big questions for the industry.

A third challenge is the fact that there are in effect currently two wind industries: one in China and one in the rest of the world. Whether the two industries can truly come together is still not obvious.

Finally, we look at what is at stake in terms of technological leadership, jobs and wealth creation for the major economies, as the EU, US and China continue to fight for domination of the wind-power industry.

Notes

1 All conversions are approximate and were calculated using the conversion rate at the time of writing.
2 The US definition of billion is used throughout, i.e. billion is a thousand million.

1 From Maoism to Lear Jets

Turbine makers go global

On September 27, 2013, at simultaneous early-morning press conferences in Copenhagen and Tokyo, officials from Vestas, the world's largest wind-turbine manufacturer, and Mitsubishi Heavy Industries (MHI), one of the world's biggest industrial conglomerates, announced a groundbreaking joint venture.

The joint venture, officials announced, was to be based around Vestas' planned V164 offshore wind turbine. With an 8MW capacity, the wind turbine's rotor diameter is bigger than the London Eye at 164m, while its height at 187m is taller than the Gherkin building in the same city. The turbine is designed to be installed in the harsh conditions of the North Sea in water depths of up to 70–80 metres and to operate for 25 years continuously.

The offshore wind business is still relatively young and there are many uncertainties. Project developers and investors prefer to deal with companies that have the financial strength to see through their warranties in case of any large-scale equipment failure. By teaming up with MHI, Vestas was bringing in a partner with a huge balance sheet – MHI revenues were over Y3trn (US$0.5trn) in 2012 – that should put it on a par – at least financially – with the undisputed market leader in offshore wind power, Siemens. MHI also brought to the table expertise and industrial capacity in areas from shipbuilding to aerospace and space technology to power plants. The joint venture virtually guaranteed that the V164 would be built and deployed on a large scale, after two years in which Vestas' capacity to bring the machine to market had looked in doubt.

However, the moment had a bittersweet feel to it. The joint venture began with a 50:50 split, but MHI has the option to increase its stake to 51 per cent in 2016, putting it in effective control. A few weeks before the announcement, Vestas had dismissed its controversial Chief Executive Ditlev Engel and replaced him with Anders Runevad, a Swedish telecommunications executive – the first time that Vestas had had a non-Danish CEO. Runevad had come, in the words of Chairman Bert Nordberg – also a Swede and also a telecoms executive – to guarantee 'a future without surprises' for Vestas.

The deal with MHI was the latest chapter in a remarkable story that has seen a group of small, mainly Danish companies sell the wind-power

Figure 1.1 Vestas' giant V164 offshore turbine (Vestas)

concept to the world and turn it into a mainstream energy source and multi-billion-dollar industry.

From Tvindkraft to the California Wind Rush

In 1972, a group of radical teachers set up a collective and base on a plot of farmland called Tvind in the small Western Jutland town of Ulfborg, around 100km away from the then headquarters of a small agricultural equipment manufacturer called Vestas. Influenced by Maoism and the current debates around radical pedagogy, its charismatic and undisputed leader was Mogens Amdi Petersen. The group began to set up a number of Tvind schools across Denmark, which were supported by Denmark's liberal public subsidies.

In 1975, the Energy Crisis had led Denmark's government to consider a large-scale nuclear programme, and political support was growing for the idea, even though public opinion was largely hostile. Next door to Denmark, Sweden was about to commence production of power from its Barsebäck nuclear plant.

Petersen and his group decided that the best way to mobilise public support against nuclear was not by protesting, but by showing that a practical alternative existed in Denmark's strong winds, and the Tvindkraft project was born. Tvindkraft foresaw the construction of a 2MW wind turbine, a size which was far bigger than anything that had been built before. The project mobilised volunteer labour and input from Danish universities.

Figure 1.2 The Tvind group's charismatic leader, Mogens Amdi Petersen (Scanpix Denmark; photographer Aage Srensen)

Surprisingly the project was a big success. Up to 100,000 people visited Tvindkraft during the three-year building work, and when it was finished in 1978, the turbine worked. The design and construction of the turbine led to a number of technical advances which influenced the future development of the industry and had a big influence on some individuals who went on to become key figures in the modern wind industry.

Henrik Stiesdal, who went on to design Vestas' first commercial turbine, built the first offshore wind project with fellow Danish company, Bonus, and is now CTO of Siemens Wind, one of the world's biggest turbine OEMs (original equipment manufacturers); he was one of the people who visited Tvind. Stiesdal says, 'During Christmas 1976 my father and I therefore went to Tvind for the first time, and like everyone else were fascinated by this group of obvious amateurs, who from something that appeared to be absolutely square one, were building the world's biggest wind turbine.' He said of his visit:

> The effect of the Tvind turbine as a source of inspiration cannot be overstated. A large number of the pioneers became hooked, like me, on the possibilities and practical challenges of wind power when they visited Tvind. The almost nonchalant self-confidence with which the so-called 'Mill Team' built something no one had ever done before was very contagious.

Figure 1.3 Siemens Wind's CTO, Henrik Stiesdal, one of the pioneers of the modern wind industry (Siemens)

Stiesdal points out that some of the tools later commonly used in the wind industry, such as blade moulds and measuring tools, were developed at Tvind. It is significant that, in keeping with the Tvind group's political philosophy, the technology for the turbine and its blade was made publicly available – making it the first 'open-source' blade design long before the term came into use.

Among others who were influenced by Tvindkraft were the young engineers at the Risø DTU National Laboratory, which was originally set up by physicist Niels Bohr to study nuclear power and would go on to be a key institution in developing Danish wind-power technology and spreading it across the globe.

Before we move on, it is worth noting that the Tvind turbine is still producing power at the time of writing. The Barsebäck nuclear plant was shut down completely in 2005.

Vestas steps in

Henrik Stiesdal and blacksmith Karl Erik Jørgensen built a 10-metre diameter turbine in 1979 through a company called Herborg Vind Kraft or HVK. After a series of experiences with storms and lost blades, they, together with blade manufacturer Erik Grove Nielsen, developed a machine with pitchable blade tips to protect the turbine from reaching too high speeds. Meanwhile Danish agricultural machinery company, Vestas, had been experimenting with wind technology – mainly in secret – since 1971, with Birger Madsen in charge of engineering.

In 1979, HVK concluded that life as a stand-alone wind-turbine manufacturer was not a realistic option. At the same time Vestas' latest turbine idea – a vertical access 'whisk'-like machine, had resulted in failure, convincing Madsen that a new approach was needed. Stiesdal and his partner agreed to license their turbine design. This move resulted in the Vestas V10 and then the 55kW V15 turbine, which became its first truly commercial machine.

Planet Zond calling Denmark

Meanwhile, on the other side of the Atlantic, the threat to the US economy from the Arab oil embargoes had led President Jimmy Carter to create the US Department of Energy in 1977, and the government began to look at ways to develop wind and solar energy.

Federally funded R&D programmes aimed at getting big defence contractors to develop wind technology were largely unsuccessful, but federal and state regulatory policies and tax credits – such as the Public Utilities Regulatory Policies Act (PURPA) – which required utilities to take power from small non-utility generators, had a huge effect on the development of wind power.

California's Governor Jerry Brown implemented a state PURPA that ensured generous payment for power for wind developers, and developers also received a generous tax break through accelerated depreciation of their assets. The policies led to the California Wind Rush. Energy writer and analyst Daniel Yergin describes it in the following way: 'Committed wind advocates, serious developers, skilled engineers and practical visionaries were joined by flimflam promoters, tax shelter salesmen and quick-buck artists. Thus was the modern wind industry born' (Yergin 2011: 595).

Developers began setting up clusters of hundreds of machines in three giant wind-rich areas – the Altamont Pass, the Tehachapi Pass and the San Gorgonio pass – only to find that they were woefully ill-equipped to stand up to the wind they found there. Many of the turbines were destroyed soon after being installed, with blades flying off and towers collapsing; and most produced far less electricity than they were expected to. Many of the wind-farm clusters became little more than eyesores.

One of the entrepreneurs most committed to wind power was James Dehlsen, who had founded Zond in 1980. Dehlsen spent New Year's Eve

Figure 1.4 US wind-power pioneer and Zond founder, Jim Dehlsen (Clipper Windpower)

that year trying to install wind turbines in a blizzard in the Tehachapi pass in order to qualify for tax credits before they expired at midnight – something which as we shall see still sounds worrying familiar in the US wind sector.

The outcome was disheartening. 'As soon as we started turning the turbines on they started disintegrating', says Dehlsen. 'The next day we picked up the pieces. We concluded we'd better get a better technology pretty damn quick' (Dehlsen, cited in Yergin 2011: 596).

Dehlsen travelled to Europe to look for turbines that were more robust. Finn Hansen, the son of Vestas' owner Peder Hansen, heard that Dehlsen was in the Netherlands and was about to buy Dutch turbines. He flew down in the company's twin-engine plane and persuaded the Americans to make the trip to Denmark, where they bought two Vestas turbines there and then.

Zond followed up with an order of 155 Vestas V-15 turbines, 'much to the astonishment of Finn and his family', according to Dehlsen; while in 1982 it ordered 550. Vestas increased its employees from 200 to 870.

Danish companies supplied 90 per cent of the turbines during the Californian wind boom, with most of them supplied by Vestas, Bonus and Nordtank. The boom culminated with orders for 3,500 units in 1985. However disaster was not far away, as the California tax credits were about to expire. Zond had ordered 1,200 turbines to be delivered by 1 December 1985. According to Vestas, 'On the second shipment, disaster strikes: the

Figure 1.5 The end of the first US wind boom left parts of California littered with dead wind turbines (Eric Horst (Creative Commons))

shipping company goes bankrupt. Anchored outside Los Angeles, Vestas misses the deadline. When the turbines finally arrive, Zond refuses to accept them – and can't even pay for the turbines already delivered' (Vestas, history 1971–86 http://www.vestas.com/en/about/profile#!from-1971–1986).

Dehlsen's description of the events of 1985 is rather different:

> We referred to that year as the D-Day of wind where the sheer logistics of multiple project locations, the number of turbines to be installed in the few remaining months of the year, with construction activities often hampered by blizzard conditions, had our team working around the clock. Thanks to my wonderfully supportive wife Deanna, who kept our field crews nourished with her chilli, we all survived the experience.
>
> (Dehlsen 2003)

In any case, by 1986 the wind rush was over, as tax credits were rolled back and oil prices fell, apparently ending the threat to US energy security. Zond was forced to cancel its planned initial public offering and negotiate with its creditor banks, going into 'full survival mode, liquidating equipment and painfully laying off a large part of our outstanding team' (Dehlsen 2003).

Back in Denmark, the government had made some – short lived – changes to the regulatory framework at the end of 1985 that were extremely unhelpful to the wind industry, requiring that people investing in cooperative wind farms had to be resident in the municipality where the wind farm was established. 'This was a kick in the back especially at Vestas where sales for Taendpipe, the biggest wind farm at that time were completed and owners were spread all over the country', says veteran wind-sector analyst, Per Krogsgaard. Vestas effectively went bankrupt and suspended its payments in October 1986. A new company, Vestas Wind Systems A/S was launched in 1987, with Johannes Poulsen as managing director.

Vestas bounces back

Interest in wind power was growing worldwide, and the company was soon manufacturing turbines for wind projects in India, backed by the Danish state-aid agency, Danida. In 1989, Vestas acquired government-controlled Danish Wind Technology, boosting its capacity and sales reach. The same year it created a German subsidiary and followed this with subsidiaries in Sweden and the US in 1992, while exports were also growing to the UK, Australia and New Zealand.

By the early 1990s, Vestas was selling turbines around the world, and was confident it could become the 'largest modern energy company in the world'. The technology on which the 'Danish concept' was built was evolving, with pitch control and lighter blades, and systems to ensure an even output of electricity to the grid. These innovations were encapsulated in its 1994 V44 turbine, which could produce 600kW of power. The company was pioneering the integration of large numbers of turbines into wind 'farms'.

In 1994, Vestas added more production capacity by buying Varde-based Volund Stalteknik. The same year it formed a joint venture with the Spanish aerospace and engineering company, Gamesa, giving it an entry into what would be one of its most important markets over the next 15 years.

In 1995, Vestas installed its first offshore turbines – the first offshore wind farm in Denmark had been installed four years earlier by fellow Danish turbine manufacturer Bonus under Henrik Stiesdal's supervision.

In 1997, Vestas built what was then the world's largest commercial wind turbine, with a 1.65MW capacity, 55 times greater than the first wind turbine produced by the company in 1979. With 32-metre long blades, the turbine could produce enough power to supply around 1,000 households.

In 1998, with its turbines representing 22 per cent of the world market, the company carried out a stock listing on the Copenhagen Stock Exchange raising €175m (US$240m). The share offer was eight times over-subscribed.

Behind the growth of Vestas was steady political support at home. As early as in its second energy plan from 1981, the Danish government had published a target of 10 per cent renewable electricity, to be met by a combination of 60,000 wind turbines and 5,000 decentralised biogas plants. Denmark had cancelled its plans for nuclear power in 1985, a year before the Chernobyl nuclear disaster. Inspired by the 1987 'Brundtland Report' – Our Common Future – Denmark began to set targets for wind-power growth as part of its overall plans to limit carbon emissions. The third Danish energy

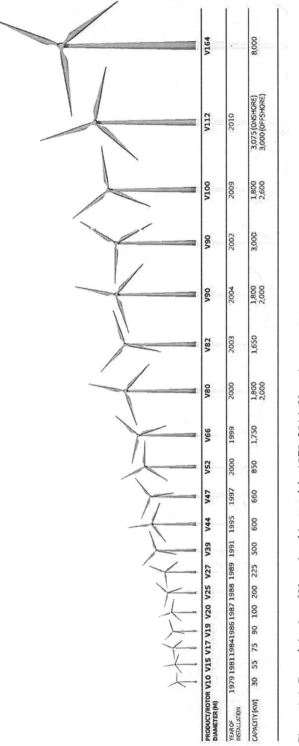

PRODUCT/ROTOR DIAMETER (M)	V10	V15	V17	V19	V20	V25	V27	V39	V44	V47	V52	V66	V80	V82	V90	V90	V100	V112	V164
YEAR OF INSTALLATION	1979	1981	1984	1986	1987	1988	1989	1991	1995	1997	2000	1999	2000	2003	2004	2002	2009	2010	
CAPACITY (KW)	30	55	75	90	100	200	225	500	600	660	850	1,750	1,800 2,000	1,650	1,800 2,000	3,000	1,800 2,600	3,075 (ONSHORE) 3,000 (OFFSHORE)	8,000

Figure 1.6 Growth in size of Vestas' turbine models, 1979–2014 (Vestas)

plan, Energi 2000, presented in 1990, called for 1,500MW of wind power to make up 10 per cent of electricity consumption by 2005 – a share that was reached in the spring of 1999. The fourth energy plan from 1996 added a long-term 2030 target of 5.5GW, of which 4GW should be installed off-shore, for wind power to meet 50 per cent of Denmark's power demand.

In the event, championed by Energy and Environment Minister Svend Auken and backed by broad political agreements on Danish energy policy, the wind industry was able to exceed these targets. Capacity grew from 400MW to 1.4GW between 1995 and 2000, when it already met 12.5 per cent of the country's power consumption. A new right-wing minority government, led by Prime Minister Anders Fogh Rasmussen, took over in the autumn of 2001, ending the long-held tradition of broad political agreements on Danish energy policies. With the support of the right-wing Danish People's Party, the coalition government had a narrow majority, which it used to dismantle the progress made under Svend Auken and former Prime Minister Poul Nyrup Rasmussen. This led to five years of relative stagnation in Denmark from 2003 to 2008.

The leader of the Conservative Party – former policeman Bendt Bendtsen, who is currently a member of the European Parliament – was given the energy portfolio in the new coalition government. In February 2002, he told Reuters that wind energy was too expensive and that other countries should be cautious about wind energy's impact on their competitiveness: 'I'm of the opinion that Denmark shouldn't continue to subsidise installation of new wind turbines after 2003.' His opinions went worldwide and the leading Danish companies found it difficult to understand why the new minister found it necessary to hurt exports, in addition to shattering the Danish home market.

'It was export masochism', says Christian Kjaer who was a policy advisor to the Danish Wind Energy Association at the time. 'It was also bizarre that the minister responsible for trade and industry warned about competitiveness at a time when Denmark, according to the World Economic Council, was the world's fourth most competitive nation' (personal communication).

A meeting held later in 2002 between Minister Bendtsen and frustrated CEOs of the Danish wind-energy companies was a farce. The new minister told the CEOs that it was not him but his larger coalition partner – The Liberal Party – that was behind the changes in policy. The CEOs left the meeting in despair, waiting for better days. They had to wait until 2007, when Connie Hedegaard was appointed Minister for Climate and Energy. She even managed to convince Prime Minister Anders Fogh Rasmussen that the coalition's policies had been a mistake. In 2008, referring to his own party's energy policies, he admitted it had been dragging its feet. 'Let me do something that I know many would like and have been waiting for me to say: that was probably wrong', he told his party's national congress in 2008 – a year before world leaders were to arrive in his capital Copenhagen to reach a global agreement on emissions reductions.

Meanwhile, momentum around limiting carbon emissions had been growing steadily outside Denmark, leading to the passing of a growing body of legislation to extend support for renewables between 1990 and the present day. Following the 'discovery' of the climate issue in the second half of the 1980s, the German Enquete commission (The Enquete Commission on Preventive Measures to Protect the Atmosphere) was established in 1987, and came out with two seminal reports, one in 1989 and one in 1990, which was the beginning of the 'Energiewende' (energy transition) in Germany. It was also was one of the first major moves in the war between the pro- and anti-nuclear forces in Germany, the latter of which had been strengthened due to Three Mile Island and more importantly, the Chernobyl disaster.

Following these reports, the first German renewables support legislation came in, the 1991 Electricity Feed in Act, the forerunner of the EEG. GWEC General Secretary, Steve Sawyer, calls the Feed in Act 'probably without much doubt the single most important piece of national legislation giving rise to the modern renewables industry, particularly wind' (personal communication).

Spain passed a landmark electricity law in 1997, setting the stage for that country's subsequent boom in wind power. But the key moment in terms of scaling up the level ambition for the renewable sector was the original EU renewable electricity directive of 2001, setting targets for all the then 15 member states of the European Union. This directive had a target of increasing the share of the EU's electricity from renewables from 14 per cent to 22 per cent by 2010. Another important development was the passing and implementation of China's Renewable Energy Law in 2005–06. 'While the drivers were initially energy security, they were enhanced throughout the 90s and the first decade of the twenty-first century by climate concerns, and the jobs and industrial development and local economic benefits drivers came along with the scaling up', says Sawyer (personal communication).

Even more compelling was the '20:20:20' EU Climate and Energy package legislation which was agreed in December 2008 and finally entered into force on 25 June 2009. 'Today tomorrow changed! The European Parliament and the Council have agreed the world's most important energy law', said EWEA Chief Executive, Christian Kjaer, on the day of the agreement.

The agreement had been preceded by a rigorous battle at the highest levels of European politics. At a meeting between European Commission President José Manuel Barroso and then president of the European Wind Energy Association (EWEA), Arthouros Zervos, Zervos had asked for sectoral renewable targets for electricity, heating and transport, because an overall binding renewable energy target seemed impossible to get through the Council of Ministers. The Commission President told Zervos that he would be able to deliver a binding overall 20 per cent target from the Member States, if the industry compromised on its demand for sectoral targets. Zervos replied that this would be acceptable for the industry, but nobody believed it possible to have all 27 governments agree on a binding target.

In January 2007, only three Member States – Germany, Sweden and Denmark – supported the idea of having a binding target for renewable energy. EU energy ministers meeting on 15 February failed to reach an agreement and left the decision to the 27 Heads of State, who met three weeks later. On 9 March, led by the German Presidency of the EU, they unanimously agreed 20 per cent binding targets for renewable energy and carbon reductions and an indicative target of 20 per cent energy efficiency, all to be reached in 2020.

'The unanimous adoption by 27 European Heads of State could not have happened at any other point in time', Christian Kjaer recalls today. 'Oil prices had just risen above \$100 per barrel for the first time, Russia and Gazprom had cut off gas supplies to Europe a few months before, the economies were booming and Germany held the Presidency of the European Union' (personal communication).

The agreement had three main elements: a carbon reduction target, a renewables target and an efficiency target. 'In terms of legislative quality, it was the good, the bad and the ugly', Kjaer says.

The renewables legislation was excellent; the greenhouse gas reduction target of 20 per cent was undermined by external credits so the effective domestic reduction was below 10 per cent, compared to the IPCC's demand for 25–40 per cent was incompatible with domestic reductions; but the really ugly one was the 20 per cent energy efficiency target, which was merely indicative and, thus, could not be enforced in Court (personal communication). Ten months later, the European Commission tabled its legislative proposals in January 2008.

During the ten months following the Heads of State's political agreement on targets, an intense battle was being fought between the European renewable energy industries, united under its umbrella organisation EREC (European Renewable Energy Council), and large industrial players within and outside the energy sector, gathered under the umbrellas of Eurelectric and Business Europe, who lobbied for the Commission to come out with a weak – or even better an outright unworkable – proposal from the Commission. The battle was also going on inside the Commission, between civil servants of opposing views.

'We were losing the battle of the Commission proposal for 9 months', Kjaer says.

> It was only the direct personal intervention by then Energy Commissioner Andris Piebalgs during the last month of drafting, that turned the tide, and a good proposal was tabled in January 2008. Claude Turmes, one of the European Parliament's best legislators, had been appointed Rapporteur on the Renewables Directive and he sealed a great final outcome of the negotiations with the Council in December 2008.
>
> (personal communication)

The 'Danish concept' goes global

By 2000, demand for wind power around the world had taken off and, led by Vestas, exports went from strength to strength. The world market had grown to almost 4GW of annual installations, of which Vestas was supplying almost a third. The 'Danish Concept' had gone global.

As different countries began to develop their wind capacity, a number of turbine companies emerged to compete with Vestas. In Spain, Gamesa Eólica was formed in 1994 to supply wind turbines and develop wind farms from an engineering group that had mainly been concentrated on aerospace. Vestas was involved, as a partner, with a 49 per cent shareholding and technology licensing agreements. The licensing agreement between Vestas and Gamesa limited the Spanish company to selling its turbines in Spain, Latin America and North Africa, but Gamesa's management had global ambitions, and there were conflicts over development activities carried out by Gamesa in places like Italy and Greece.

In 2001, Vestas sold its stake in Gamesa for €287m (US$395m). Vestas' CEO Johannes Poulsen said at the time:

> Differences in strategy between Vestas and Gamesa have led to an increasing number of strategic conflicts in the marketplace. We have therefore been looking for ways and means to avoid this and our sale of Gamesa Eólica shares seems to be the better option for all parties involved.
>
> (*Windpower Monthly* 2001)

Under the terms of the agreement, Gamesa was able to continue using Vestas' technology, including for the new V80-2MW turbine, while Vestas gained access to the fast-growing Spanish market. 'We are creating ourselves a new competitor for whom we have the highest respect as we know his industrial capability as well as his basic technology', said Poulsen. 'But we prefer this over a situation that creates uncertainty about Vestas strategy and we believe the market for wind power gives room for both of us in the future' (*Windpower Monthly* 2001).

Gamesa went on to sell its aerospace division in 2006, and in the same year was in second place in the global wind-turbine supplier rankings with over 10GW installed and a market share of over 15 per cent.

In Germany, engineer Aloys Wobben had established Enercon in 1984 and built the company's first turbine, the 55kW E-15/16. In 1993, Enercon produced its E-40/500kW series using a highly original direct-drive configuration with an annular generator that eliminated the need for a gearbox, which had historically been one of the most common areas of mechanical failure. Enercon turbines proved to be extremely successful among the small developers who are the driving force in Germany's wind market, due to its reputation for reliability and its highly popular PartnerKonzept service model. Enercon's attempts to expand internationally after 1996 were less consistent than those of Vestas and Gamesa, and Wobben found himself embroiled in a number of patent disputes with rivals such as Vestas and Kenetech/Zond that hampered operations, particularly in the US. However, the company built up operations in Sweden, Brazil, Canada, Portugal, France and Austria. It has consistently been in fourth or fifth place in the global supplier rankings, helped by its dominant position in Germany.

Wind-sector consultant and analyst Aris Karcanias notes of Enercon: 'They have managed to consistently maintain a leading position amongst the top-five turbine OEMs despite not competing in the world's two largest markets, the US and China, and actively deciding not to develop turbines for the off-shore market' (personal communication). He adds that Enercon is considered by many to be the 'Apple' of the wind industry because of its pioneering of a direct-drive wind turbine without the use of permanent magnets, which proved to be a pinch point for other OEMs in later years, as well as concrete towers.

As we will see, as wind markets developed outside Europe, they developed their own national champions, as in the case of India's Suzlon, which was founded in 1995 and expanded rapidly thereafter, and China's Goldwind, founded in 1998.

Ditlev Engel and 'The Will to Win'

In early 2004, Vestas took over NEG Micon, the second-largest Danish wind company. The merger allowed Vestas to achieve a global market share of 34 per cent and revenues for the year of €2.56bn (US$3.53bn). Commentators saw the mergers as a defensive move by Vestas that would prevent NEG Micon from being taken over by a company from outside Denmark, including Gamesa, which was making increasingly aggressive moves in the market after its split from Vestas a few years earlier. Critics said that Vestas had overpaid for the poorly performing NEG Micon, calling the deal a rescue rather than a merger. The merger deal had been under discussion for over a year, but NEG Micon reportedly 'surrendered' to a takeover by Vestas after it was forced to issue a profit warning in November 2003 (*Windpower Monthly* 2004). Managers of the combined company said they were targeting savings of €67m (US$92m) a year through the reduction of overlapping sales and production facilities as well as better deals with suppliers of components and raw materials.

In October the same year, Vestas hired Ditlev Engel as its new President and CEO to replace Svend Sigaard. Engel had worked his way through the ranks – and through business school – at Danish paint manufacturer Hempel, and become the company's CEO in 2000 at the age of 35. Without any knowledge of wind power and coming from outside the engineering fraternity that dominates the sector, Engel says he spent the two months before he took up his new post in May 2005 travelling the world and getting to know the business.

Figure 1.7 Ditlev Engel, who presided over wholesale expansion at Vestas after taking over as CEO in 2005 (Vestas)

On 26 May, he presented the company's first-quarter results along with a new strategic plan with the Nietzschean name 'The Will to Win'. Key to the vision was a plan that would supposedly take wind power from being a still relatively marginal source of energy to one that was on a par with oil and gas.

'Many people still regard wind power and thereby Vestas as a "romantic flirt" with alternative energy sources. It is not. Vestas and wind power is a real and very competitive alternative to oil and gas', said Engel.

Engel's strategy promised, among other things, that Vestas would achieve:

- earnings before interest and tax (EBIT) of at least 10 per cent by 2008 from an expected 4 per cent in 2005
- a global market share of 35 per cent
- a reduction in production costs including a staff reduction.

Along with the new targets and vision, Engel created three new business units to add to Vestas' expanding corporate structure – they were Technology, People and Culture and Offshore, which were added to Nacelles, Blades, Northern Europe, Towers, Mediterranean, Asia-Pacific and Americas – while centralising decision making in an Executive Board of Management that had two members, himself and Executive Vice-President and CFO, Henrik Nørremark. As we shall see, the relationship with Nørremark, who unlike Engel was a Vestas veteran, would play a fundamental role in Engel's career in Vestas.

In language that was part business management speak and partly a throwback to the heroic discourse surrounding the Tvindkraft project, employees were urged to increase their efforts, with Engel saying 'This work will no doubt be exciting and very hard. At the same time, it will require the will to change for all of us.' At the same time a sculpture called 'The Willpower' was commissioned from artist Jørgen Pedersen and copies were installed outside Vestas' offices around the world (see Figure 1.8). Corporate literature described the significance of the sculpture in the following terms:

> Vestas was founded to take on one of the biggest challenges facing the earth. We have never been, and will never be, characterised by losing faith and commitment when met with resistance and inertia. On the contrary, Vestas and our employees have shown persistence in breaking down misconceptions towards wind energy and supporting the growth of the industry. We call this our Willpower. It is expressed in the sculpture entitled 'The Willpower', which is placed at a number of the Group's locations. Reaching for the sky, it symbolises the willpower and passion possessed by the employees.

For the next eight years, Engel would become the best-known face of the wind industry. As well as becoming a celebrity in Denmark, he appeared in big financial forums such as the World Economic Forum in Davos, took part in working groups run by the United Nations and the IEA, and was a regular on business news channels like CNBC, CNN and Bloomberg TV. There is no doubt that he made a big contribution to raising the profile of the wind industry and getting it taken seriously in the wider world of politics and business as a real alternative to fossil-fuel power generation. Engel kept a frenetic timetable, flying by private jet from country to country. Quarterly financial results were held alternately in London and New York, for instance.

Engel's leadership was both dynamic and charismatic. However, the consensus is that it was not successful in terms of reaching the objectives that he had set out in 'The Will to Win'.

Despite talking initially of a reduction in staffing, Engel presided over a more than doubling of Vestas' head count, which reached a high point of

Figure 1.8 'The Willpower' sculptures were installed outside Vestas' main offices under Engel (Vestas)

23,252 in 2010. Vestas opened its first factory in China in 2006, and its US blade, tower and nacelle plants in 2008–10. It added several more factories in European locations including the UK, Italy, Spain and Germany, along with a network of R&D centres.

Revenues and annual orders soared as the company tried to grow at full speed to accompany the growth of the wind market. Engel did reach the target of a 10 per cent EBIT margin in 2008, the year when Vestas' profits hit their peak and wind was effectively a 'seller's market', with demand for turbines outstripping global manufacturing capacity. But, as the rate of market growth – outside China – began to slow, capacity outstripped demand and prices and margins began to fall again.

Meanwhile Engel's ambition of maintaining a market share of 35 per cent proved impossible as competition intensified. Vestas found itself being squeezed, particularly in China by fast-expanding local producers, and in

Table 1.1 Vestas' key financial indicators

Year	Revenues (€ m)	Ebit margin (%)	Net profit/(loss) (€ m)
2004	2,363	(2.1)	(61)
2005	3,583	(3.2)	(192)
2006	3,854	4.9	113
2007	4,861	5.3	104
2008	5,904	10.4	470
2009	5,709	4.9	125
2010	6,920	4.5	156
2011	5,836	(1.0)	(166)
2012	7,216	(9.7)	(963)

Source: Vestas. Figures were revised after Vestas changed its accounting procedures in 2010

Table 1.2 Vestas' turbine deliveries

Year	Deliveries (MW)
2004	2,784
2005	3,185
2006	4,239
2007	4,502
2008	6,160
2009	6,131
2010	4,057
2011	5,054
2012	6,171

Source: Vestas

the US, where there was intense competition from both fellow European producers and from GE. Vestas' ambitions to maintain its market share were an important factor behind a deterioration in margins on turbine sales, as the company entered into contracts at unfavourable prices between 2009 and 2011.

On 27 October 2009, on the eve of the Copenhagen climate talks, Engel announced new, even more ambitious targets for Vestas; the so called 'Triple 15'. These foresaw achieving an EBIT margin of 15 per cent and revenues of €15bn (US$21bn) no later than 2015. As the company literature noted, this translated 'into an average annual growth of at least 15 per cent and a substantial improvement in the EBIT margin'.

'Vestas' triple 15 goals looked ambitious when they were announced, as there were already signs of market weakness starting to appear then', says analyst Robert Clover (personal communication). He notes that equity markets never seemed to believe the targets would be met with analyst consensus forecasts 'way below' Vestas' margin and sales targets. 'Had the market

Table 1.3 Vestas' share of the global market (2005–12)

Year	Market share (%)
2004	34
2005	27.9
2006	28.2
2007	22.6
2008	19.8
2009	12.5
2010	14.9
2011	12.9
2012	14

Source: BTM-Navigant

believed triple 15, then the stock would have been valued in excess of DKK 1,000', Clover says (personal communication).

Just a few months later, with the failure of the Copenhagen talks and demand growth beginning to slow as the financial crisis hit home, it started to become clear that Triple 15 had been a mistake, and in 2010 Vestas saw its first significant drop in deliveries since the aftermath of the California Wind Rush.

The wholesale expansion which Vestas carried out under Ditlev Engel was understandable, given the speed of growth of the global wind market in 2005–09, but it had left the company overexposed. Most analysts are also highly critical of Vestas' control over costs during this period. As we noticed, Engel had added more business divisions to the company's already complex structure, adding new layers of middle management, with many of the divisions acting as complete companies in their own right. And, as would become apparent in 2011 and 2012, there were serious lapses in the company's cost calculations and the execution of its new technology programmes.

All this left Vestas vulnerable, and paved the way for a series of profit warnings in 2011 and early 2012 that shocked investors, and a rapid deterioration in the company's finances. By 2012, the EBIT targets of 10 per cent and then 15 per cent had turned into a negative margin of 9.7 per cent and a loss of almost €1bn (US$1.4bn). The main factor behind Vestas' growing problems was increased competition. With the wind market gaining momentum throughout the decade and growing at a rate that surprised even its most enthusiastic advocates, and returns being high, Vestas and the other pioneering companies were not going to have the wind-power market to themselves for long, as some of the world's biggest industrial companies made their moves.

We will look at how Vestas managed to survive the crisis in Chapter 7.

2 Big industry moves in

GE sweeps up Zond

As we have seen, James Dehlsen was left licking his wounds at the end of the California Wind Rush. The US wind industry had gone into a deep decline, with most of the companies going bust.

Dehlsen had the advantage that Zond had shares in all the projects he had developed, and he spent the next years re-engineering the California projects that had already been built; and says he succeeded in raising output by 22 per cent between 1986 and 1989. He also persuaded utility Florida Power and Light (FPL) to invest in Zond's remaining projects in the Tehachapi Valley, leading to the building of the 77MW Sky River facility, which at the time was the largest single project in the US.

Dehlsen says that surmounting major obstacles – including the demand from utility Southern California Edison that the developers provide 75 miles of 220kV transmission lines, adding US$30 million to the project cost – and completing Sky River meant that 'Zond's survival was now assured, and we could finally make good to the banks and our other generously supportive creditors who had grown to share our vision' (Dehlsen 2003).

The 1991 Gulf War and the Energy Policy Act of 1992 had seen the reintroduction of tax credits for wind power, but with the important difference that the new credits rewarded the actual production of electricity from wind projects and not just investment in building new turbines. As the 1990s went on, different states began to implement renewable portfolio standards, and the US wind industry began to pick up speed again.

The question for Zond was now how to build its own turbine to lower the cost of energy and allow it to capture the manufacturer's profit margin, as Dehlsen saw it. The company received support from the National Renewable Energy Laboratory (NREL) and the DOE in 1993 for work on Zond's 550kW Z40 machine, which was completed in 1995. Zond then acquired patent rights from the US's biggest and best-known turbine manufacturer, Kennetech, when it went bankrupt in 1996, as Dehlsen pursued variable-speed technology. Zond produced a 750kW turbine and then started work on a 1.5MW turbine.

Dehlsen says he approached the then growing energy giant Enron in 'anticipation of the company's capital needs for project development and manufacturing ramp-up' (Dehlsen 2003) and sold Zond for US$100m. Enron officials say that they had decided to make a play on wind energy, and 'brought Zond back from the brink' (former Enron official Robert Kelly, cited in Yergin 2011: 601). In any case, on Dehlsen's urging, Enron then acquired the bankrupt Germany turbine manufacturer Tacke, and by combining technology from the two companies, Zond produced the TZ 1.5MW turbine. Although Enron Wind's sales soared between 1997 and 2001, in autumn 2001 Enron collapsed spectacularly into bankruptcy and scandal.

Industrial giant General Electric had been considering getting into the wind business for several years. Two key executives – Mark Little and Jim Lyons – unsuccessfully pitched the idea of a major move into the sector in the late 1990s under former CEO Jack Welch. 'When Jack was CEO we did not get a warm reception', says Little. 'He did not see wind as a serious business' (Magee 2009: 50).

Just ahead of the Enron bankruptcy, however, a dynamic new CEO, Jeff Immelt, had taken over the leadership of GE. While cautious about the high cost of entry into the wind business, Immelt had already understood the long-term strategic potential of the wind industry.

In 2002, GE stepped in and bought Enron's wind business for US$328m, creating a new GE Wind Division. In its second-quarter earnings report of 2002, GE said that the deal 'established GE Wind Energy as a leading presence in the renewable-energy industry, which is growing at nearly 20% a year'. Note though that GE subsequently went to a court to ask for a refund of 50 per cent of the purchase price (see Mumma 2002).

While the technological elements were there, GE had to invest heavily and bring a lot of engineering expertise to bear to create a 1.5MW turbine that would meet its strict standards. 'The industry was fundamentally broken', said the former head of GE Wind, Vic Abate. 'We need to go through and rigorously re-engineer to make it a mainstream technology.' Immelt goes even further. '[Enron's wind business] was a very broken model that took us three years to fix', he says (Magee 2009: 54). When the task was completed, however, GE had a turbine that would prove to be one of the best-selling models in the world, one which would make a serious dent in the sales of the Danish incumbents.

By 2005, the US wind industry was picking up speed fast, with the annual installed capacity growing at an average rate of 40 per cent between 2005 and 2009. In 2009, GE's 1.5MW machine made up half of the entire wind-turbine fleet in the US, with 10,000 units installed. It followed the 1.5MW machine with the larger GE2.5XL, which was installed in the giant 845MW Shepherds Flat wind farm in 2011. GE also developed a 4.1MW offshore turbine, after acquiring Norwegian turbine developer ScanWind in 2009 and installed an offshore prototype in Gothenburg, Sweden at the end of 2011.

Figure 2.1 GE turbines at a site in the US (GE)

GE's wind division has consistently been in the top four turbine manu-facturers, helped by its global financing and service capabilities. In 2012, after a record year that saw around 13.2GW installed in the US, analysts at BTM-Consult and IHS-Emerging Energy Research estimated that GE had knocked Vestas off its number one spot, with a global market share of 15.5 per cent.

On the other hand, it is worth noting that GE's ability to win market share outside the US has been relatively limited, although it has done well in individual markets such as Turkey and Brazil. In 2012, 75 per cent of the turbines GE delivered globally were in the US, where it commanded a 38 per cent share of the market. By contrast, in Europe its market share sat at about 6 per cent, trailing its global rivals. Three of them – Vestas, Siemens and Gamesa – all held far higher market shares in the US last year than GE did in Europe.

In 2013, Anne McEntee, a veteran of GE's oil and gas business, replaced Abate as the head of GE's renewables business. GE also appointed Cliff Harris as its new general manager for renewables across Europe, the Middle East and Africa with the task of raising the company's market share. Harris told *Recharge* that he has a personal target of doubling GE's European mar-ket share by 2015. Time will tell whether GE has the advantages to be able to do this (Stromsta 2012).

Siemens Wind goes for scale

We have already met Henrik Stiesdal at the Tvindkraft project and in the earliest development of the 'Danish concept' and Vestas' first viable turbine. After working as a consultant at Vestas and then taking time off to complete his studies, Stiesdal joined Danish wind-turbine manufacturer Bonus Energy in 1987 as first a development specialist and from 1988 technical manager, becoming Chief Technology Officer in 2004.

In 1990–91, Stiesdal was responsible for building the world's first off-shore wind farm at Vindeby, Denmark, with 11 specially adapted 450kW turbines. In 1998, he designed Bonus's first variable-speed turbine and introduced a series of innovations in blade manufacturing and control.

In 2004, the Bonus management sold the company to German industrial giant Siemens for an undisclosed amount that was estimated at anything between US$240 and US$400m, with Stiesdal staying on as CTO. The company maintained its manufacturing operations in Brande, Denmark. Analysts said that Siemens had made the move after closely watching GE, and in the fear that it would lose out to its US rival in a strategic growth area. Like GE, Siemens was looking to compensate for an expected fall-off in orders for conventional power-generation equipment, like gas and steam turbine plants, with rising sales in wind equipment.

Andreas Nauen, then vice-president of communications and strategy at Siemens Power Generation (and later to lead first Siemens Wind Energy and then REpower), said: 'We wanted to find a company that has the same business approach as us and that would complement our business.' Nauen also noted that 'Bonus has a superb reputation, is very risk aware and its technology is a good fit' (quoted in Hoel 2004). Siemens was particularly interested in building larger turbine models, and Bonus was already testing a 3.6MW turbine design for offshore, Nauen pointed out.

Following the Bonus acquisition, Siemens betted big on the still small but potentially massive offshore wind sector (see Chapter 5). It was able to avoid some of the highly costly failures offshore that Vestas suffered, and its 2.3MW and then 3.6MW offshore turbines became highly regarded. Having the solid financial backing of a company of Siemens' size was also a major advantage; by 2009, the company had a market share of nearly 75 per cent of the installed offshore market, rising to nearly 85 per cent by 2012.

Bonus had been a much smaller company than Vestas or NEG Micon, but also a more profitable one. Onshore, Siemens grew more slowly, with a step-by-step ramp-up of its production facilities and a more cautious approach than Vestas when deciding to set up facilities in new markets. However, by 2012 it had jumped into third place in the global rankings, with 9.5 per cent of total installations, according to BTM-Consult.

Wind power grew from 0.5 per cent to 5 per cent of Siemens' turnover between 2004 and 2011, with Siemens Wind increasing its employees from

Figure 2.2 Siemens' SWT3.6 has been the best-selling offshore turbine in the world (Siemens)

800 to the current level of around 8,000. Most of its production is still carried out in Brande, but it also set up facilities in the US in 2009 and 2010, as well as blade and nacelle factories in China and a blade-manufacturing facility in Canada. As we shall see, it also set up a major joint venture with Shanghai Electric to facilitate its entry into China's market and made a final investment decision to set up a major new factory to produce offshore wind turbines in the UK in March 2014.

In mid-2008, Siemens began testing direct-drive turbines and in 2009, it installed a 3MW prototype near Brande. This was followed by a 2.3MW version and a giant 6MW offshore direct-drive machine, with the first offshore prototypes deployed in DONG Energy's Gunfleet Sands project in the UK in 2013.

In 2011, the Renewables Division of Siemens Energy was split and Siemens Wind Power became a separate division with headquarters in Hamburg, Germany. The majority of offshore and a significant part of onshore competences remain in Denmark, and Stiesdal is still CTO, giving the company remarkable continuity in terms of technological development and understanding of the wind sector's evolution. Of the current players, Siemens is a good bet to be one of the top three, or perhaps even the top wind-turbine company, in the coming years.

Alstom and Ecotècnia

French engineering giant Alstom has long been a heavyweight in hydro-electric power-generation equipment, as well as transport, power transmission and other areas. In mid-2007, it acquired Spanish turbine manufacturer Ecotècnia for €350m (US$480m).

Founded in 1981 and with its headquarters in Barcelona, Ecotècnia was one of the leading Spanish turbine manufacturers along with Gamesa and Acciona Windpower and it had a global market share of around 2 per cent, with most of its installations in Spain. Its turbines were based on a robust design meant for the turbulent winds of mountainous Spanish areas, which isolated the gearbox from the main drive, reducing non-torque gearbox loads, and a modular construction that was intended to facilitate construction in hard to access areas. These designs became the basis for Alstom's turbine portfolio, and the company scaled up the size of its machines and blades, producing the first ECO100 3MW turbine in 2009. The company was renamed Alstom Wind in 2010. The same year, Alstom began constructing a turbine nacelle factory in Amarillo, Texas, to add to its assembly plants in Somozas and Buñuel in Spain.

As well as winning some major onshore orders in Europe, such as the second phase of Iberdrola's giant Whitelee wind farm in Scotland, Alstom Wind's main success has been in Brazil, where it has been the fastest growing turbine company in recent years (see Chapter 4).

Alstom has also moved into offshore, with its 6MW direct-drive Haliade machine, which it began testing in early 2012. Alstom has yet to register in the ranking of the top ten turbine manufacturers and the company seems to be taking a cautious approach to entering new markets. Officials maintained throughout 2012 and 2013 that the company had the ambition to be one of the top three or four players in the world and its experience in Brazil showed a willingness to make aggressive moves when it saw that the opportunity is there. However, at the time this book was going to press, Alstom's management subsequently decided that the demands of becoming a world leader in energy were too much for the company's balance sheet. A proposed deal to sell the energy assets to GE led to Siemens also launching a takeover bid, in a battle that was being resolved when this book was being produced.

Offshore wind, with its very big demands in terms of capital expenditure and warranty risk has shown itself to be an opportunity only for companies with strong nerves and balance sheets. As well as Alstom, offshore wind has attracted the entry of French nuclear energy company Areva, and the South Korean chaebol Samsung and Hyundai. Japan's Mitsubishi (now as part of

Figure 2.3 Alstom's 6MW Haliade offshore turbine being erected (Alstom)

a joint venture with Vestas) and Hitachi are also currently both making a push into offshore wind.

Life gets harder for the 'pure players'

The outcome of the entry into the market of big industrial companies like GE, Siemens and Alstom has been to make life harder for the 'pure play' wind-turbine producers such as Vestas, Gamesa and Suzlon. The diversified groups can often count on ongoing relationships with big utilities, as well as the ability to offer finance through other group divisions, but the key factor is bigger balance sheets to cope with any claims on warranties due to turbine failure, particularly in the offshore arena. Arguably, the diversified companies are able to cope better when there are lean years in the wind sector, because they are often naturally 'hedged' through their involvement in other energy-generating technology.

In short, a couple of bad years in the wind sector are not going to have the same effect on Siemens or GE as they will on Vestas or Gamesa, although it is important not to exaggerate this point, as parent companies have shown that they will not tolerate consistent losses from their renewable energy divisions – as Siemens proved when it shut down its loss-making solar business in 2013. Some of the diversified companies can also bring considerable

R&D resources into play as well, and leverage their industrial set-ups. GE is the most obvious example, with its research labs working on everything from fabric blades to better storage and 'intelligent wind farms', while it has also been able to divert staff and facilities from making gas turbines to making wind turbines during boom years, before re-assigning them to their original duties.

Wind development scales up as utilities come in

One factor that was important in attracting large-scale industrial companies to the wind market was the increasing scale of wind developers, as a series of large-scale utilities and power companies entered the sector. The increasing size of projects and investors meant that there were increasing opportunities for large-scale contract wins rather than large numbers of small turbine sales. The entry of utilities into the business also meant that the likes of GE, Siemens and Alstom were talking to the same companies that they were selling to in their other power businesses.

The rise of very large-scale wind developers was led initially by a small group of companies in Europe and the US. The most notable is Spanish utility Iberdrola, which formed its Iberenova business in 2001 and built it – through a mixture of organic growth and acquisitions – into the largest wind-power company in the world with a capacity of over 7GW by 2007, and a growing presence in the Americas and Europe. That same year, Iberdrola floated Renovables on the Madrid stock exchange, raising around €4.5bn (US$6.2bn).

In the US, utility Florida Power and Light was an early mover into wind power, despite there being no wind projects in Florida. The company had invested in wind projects as part of a diversification programme in the 1980s, and found itself owning a number of wind farms when a number of these projects went bankrupt. The company, now known as NextEra Energy Resources, stuck with the wind business throughout the 1990s and is now the biggest developer in the US – accounting for around 20 per cent of total wind generation – and the second biggest in the world, with 10.21GW of installed wind capacity as of 31 December 2013.

In Europe, Iberdrola was among a group of European utilities that built their renewable energy arms into dynamic operations and in many cases floated renewables subsidiaries on the stock market. These include Portugal's EDP (EDP Renováveis); Italy's Enel (Enel Green Power); France's EDF (EDF Energies Nouvelles) and GDF Suez; Germany's E.ON (E.ON Climate & Renewables); as well as the Spanish infrastructure company Acciona, which built Acciona Energia into a massive pure-play renewables power producer. Also notable are Denmark's DONG Energy, which has become the world's first utility whose development is centred on offshore wind; and Warren Buffet's MidAmerican Energy Holdings, which has become a major US wind player in recent years.

Figure 2.4 Iberdrola's CEO, Ignacio Sanchez Galán, transformed the Spanish utility into the biggest wind-power operator in the world (Iberdrola)

Table 2.1 Iberdrola's installed renewables capacity

Year	Capacity (GW)
2007	7,098
2008	9,302
2009	10,752
2010	12,532
2011	13,690
2012	14,034
2013	14,247

Table 2.2 NextEra's wind-power installations

Year	Capacity (GW)
2008	5,388
2009	6,493
2010	7,624
2011	8,386
2012	8,887
2013	10,117

These companies have been joined in recent years by a group of Chinese generators which include China Lonyuang Power Group, China Huaneng Group, Datang, China Huadian Corporation, China Guodian Corporation and China General Nuclear Wind Power. These Chinese companies have pushed into the ranking of the world's top developers, occupying four out of the top ten positions and ten out of the top twenty-five spots.

In 2012, the largest global wind owners accounted for 117.4GW, according to analyst IHS-EER, or fully 43 per cent of cumulative global installed capacity. The effect that the development of large-scale wind-based utilities had on the industry cannot be underestimated. Wind projects had been increasing progressively in scale throughout the 1990s. Utilities, with their ability to finance projects on their own balance sheets, could increase the size of wind farms to an extent few had hitherto imagined.

As well as being able to finance wind farms, the big power producers also had significant capacities in project management and often in engineering. Scale and the ability to plan a pipeline of projects over a relatively long period meant that the large power producers were able to sign multi-year framework agreements with turbine suppliers, with contracts foreseeing often over 1GW in capacity.

The increasing size of these contracts allowed big turbine producers such as Vestas, Gamesa and Siemens a higher degree of visibility in terms of revenues, and allowed them to ramp up investments and production. Examples include the 1.15GW deal that Siemens signed with German utility E.ON in 2008; the 2.1GW framework agreement that Vestas signed with Portugal's EDPR in 2010; agreements between Enel Green Power and both Siemens and Vestas for a total of 2.6GW; and two overlapping framework deals signed between Gamesa and Iberdrola in 2008 and 2011. Denmark's DONG Energy signed the biggest-ever offshore framework in 2009, for a total of 2.07GW of Siemens turbines.

In turn, the increasing scale of power producers contributed to bringing down the cost of energy (CoE) of wind in a number of ways. First, their investment helped push up the average size of wind farms, which studies suggest reduces cost – at least to a certain point. Second, the big producers were able to use their purchasing power to push down or at least limit turbine prices. And third, by establishing control centres that processed raw performance data from turbines – something which had previously been the jealously guarded preserve of OEMs – the large producers were able to introduce a higher degree of transparency into the market.

At Iberdrola's CORE (Renewable Energies Operation Centre) in Toledo, staff sit watching giant monitors, where they can monitor and control wind farms around the world, including being able to look 'inside' individual wind turbines. CORE director Gustavo Moreno Gutiérrez told me during a visit that Iberdrola had worked hard to break down a culture of secrecy among wind-turbine manufacturers – Iberdrola uses at least nine different ones – and ensure that the company has access to all the information it

Figure 2.5 Iberdrola's CORE wind-farm control centre (Iberdrola)

needs to be able to ensure cost efficiency. 'We can compare all kinds of performance data and that allows us to push efficiency standards higher', he said, adding that until 'very recently', only the manufacturers 'had the key to turbines and the operators couldn't even get in'.

Negotiating with OEMs to make sure that access to the right information was provided was one part of the challenge, but dealing with the data was another. 'One of the difficult things was to get the right type of information and then integrate all the different types of data from different manufacturers using a common standard', says Moreno.

Being able to compare availability across turbine types means that performance gaps between different models and turbines of the same type narrowed noticeably, say officials from Iberdrola. Other companies, such as EDPR, report similar results.

The leading utility developers achieved spectacular growth for their portfolios during 2005–11, with the two Spanish developers Iberdrola and Acciona, NextEra in the US, and Portuguese-owned but Madrid-based EDPR in the forefront and others such as E.ON investing heavily. Siemens' Henrik Stiesdal points out that during the period when the financial crisis limited the flow of investment to smaller developers, the utilities were able to keep market growth going by financing wind farms from their own balance sheets.

In offshore, utilities played an even more fundamental role in building the sector, with companies like DONG Energy, E.ON, RWE, SSE and Vattenfall taking the lead. 'I am convinced that virtually none of the offshore capacity we have seen built over the last five years would have been built without the involvement of utilities', says analyst Robert Clover (personal communication).

According to Clover, utility/IPPs (independent power producers) account for around 35 per cent of wind-asset ownership in Europe – where smaller developers and community wind farms are prevalent – but a much larger share of new investments. In China, utilities and IPPs account for more than 90 per cent of asset ownership, while in the US they account for 75 per cent.

After the big growth spurt of 2005–11, growth began to slow for the Western utilities, as a number of factors like a lack of availability of power purchasing agreements (PPAs) in the US and the Spanish crisis hit home. Attempts to make up for the slowdown by expanding into new fast-growing frontier regions in Eastern Europe hit regulatory and infrastructure barriers. An example is Romania, where Iberdrola announced its huge 1.5GW Project Dobrogea – hailed as the largest onshore wind project in the world – in 2010. Iberdrola CEO Ignacio Sanchez Galán flew to Bucharest in September 2010 to meet with Romanian Economy Minister Ion Ariton in an attempt to ensure that the project obtained grid access, after receiving a permit from transmission operator Transelectrica a few months earlier (Backwell 2010a).

Three years later, after a series of disputes over land with Czech utility CEZ – a rival wind developer – and faced with seemingly insurmountable bureaucratic barriers to establishing the transmission infrastructure it needed, the window of opportunity had passed, as Romania realised it had built too fast at too high prices. In October 2012, after completing just one 80MW project in the country, Iberdrola CEO Ignacio Sanchez Galan effectively called time on Project Dobrogea.

The publicly floated vehicles that the utilities had established had suffered from a lack of investor appetite, particularly after Copenhagen, and their share prices consistently failed to reflect the value of their assets or development portfolios. This meant that, on the whole, they did not serve their purpose as vehicles for bringing additional investment into the sector, and Iberdrola and EDF both moved to reincorporate their subsidiaries in 2011. In contrast, however, Italian utility giant Enel's Enel Green Power (EGP) subsidiary has thrived since being spun off in 2010, with a self-financing model based on steadily growing cash flow.

Stagnant or falling power demand in Europe, the effect on utilities' 'legacy' fossil-fuel generation assets of increasing quantities of renewable power in the grid, and big capital investment requirements driven by shut-downs of existing fossil and nuclear capacity put pressure on company balance sheets and brought debt downgrades from credit rating agencies.

Utility executives became increasingly strident in their appeals for support for their fossil-fuel operations, arguing that gas capacity in particular had to be supported to provide back-up for 'intermittent' wind and solar supply, as well as criticising solar PV generation in particular as high-cost. Even when continuing to invest in renewables, and with their renewables businesses often the most profitable among their divisions, utility bosses began to call for governments to slow down the development of wind and solar or even tax renewables.

In October 2013, the CEOs of ten European utilities, including big wind investors like Iberdrola, E.ON and Enel, held a joint press conference in Brussels where they claimed that renewables were wrecking the European power industry.

'European energy policy has run into the wall', GDF Suez CEO Gerard Mestrallet said. 'In sectors like steel, cars and refining, when there was over-capacity, capacity was closed. But in the energy sector, we have massively subsidised additional capacity in solar and wind, which has led us to the absurd situation in which we find ourselves today', Mestrallet said (De Clercq 2013).

Fulvio Conti, CEO of Enel, said the European industry 'needs to become investable again'. He claimed that 'Too many subsidies are being given to intermittent renewables ... and to local coal', which he said was creating a 'disaster' (Lee 2013a).

As well as a reining-in of subsidies for renewables, the CEOs appealed for an EU-wide system that rewards back-up capacity, a reform of Europe's struggling carbon-trading system and the setting of 2030 emissions targets to underpin de-carbonisation. The CEOs seemed to have had at least partial success in lobbying for this agenda in the European Commission white paper that was published in January 2013. The white paper calls for an EU emissions target of 40 per cent by 2030, and a renewables target of a modest 27 per cent, which is furthermore not binding on individual EU countries.

However, this had not been agreed by member countries at the time of writing. It is still too early to say how the conflict of interest within European utilities will work out, and how they will end up lining up in debates on energy regulation in the coming years. Many of the utilities represented in Brussels have thriving renewables businesses, whereas others are more reliant on their traditional fossil and nuclear power businesses.

There are also legitimate concerns amid the complaints about renewables. Dysfunctional power markets in Europe have had the perverse effect of pushing gas power generation off grid when the wind is blowing, while allowing an increased use of much dirtier coal (see Nicola and Bauerova 2014).

Certainly not all of the current protests from utilities about Europe's energy regulations come from the point of view of an unenlightened defence of the continent's legacy generation system. Michael Lewis, Chief Operating Officer for Wind at E.ON, the German utility that has been one of the most active investors in the sector, told a forum I organised in Frankfurt:

It's not that utilities – and certainly not E.ON – are against renewables. What we are saying is yes we support renewables, but we support them in the context of a market that functions properly and keeps the lights on. And that bit of the message is not getting through.

(Remarks at *Recharge* Thought Leaders event, EWEA Offshore Conference 2013)

Ultimately, utilities will take their lead from politicians and regulators. Clear rules to progressively decarbonise electricity generation and unequivocal political signals that there will be no turning back will allow those utilities that moved first and decisively into renewables to successfully make a transition to a new business model. Others, however, particularly those most tied to dirty, coal-based generation, show every sign of wanting to sacrifice the struggle to restrict climate change to safe levels for their own narrow interests. Given their current balance-sheet problems, utilities may not be as dominant in Europe as they have been, but they will continue to be the biggest players, particularly in the offshore arena.

Like wind-turbine manufacturers, financing constraints have created considerable scope for consolidation in the utility–developer space, and led companies to seek investment from capital-rich Asia. One of the most notable signs of the changing landscape was sale of a 21.35 per cent stake in Portuguese utility EDP to China's Three Gorges China Corporation for €2.7bn (US$3.7bn). Three Gorges manages a huge portfolio of hydropower assets (including the mega-project of the same name), but it is also a major investor in wind farms, and an investor in leading Chinese wind-turbine manufacturer, Goldwind.

Under the terms of the deal, Three Gorges agreed to invest €2bn (US$2.8bn) in a portfolio of 900MW of operational and 600MW of ready-to-build projects by EDP Renovaveis between 2012 and 2015. Of these, 600–800MW will be in Europe, a similar amount in the US and up to 200MW in South America. The first €800m (US$1,000m) will be invested during the first 12 months after closing the agreement. 'We are going to combine efforts to become worldwide leaders in renewable energy through joint development and ownership of selected renewables projects', EDP CEO António Mexia says (Mexia 2011).

CTG Vice-President Lin Chuxue told reporters in Portual: 'EDPR is an excellent platform for the future for Three Gorges', adding it will help the group 'to arrive at other businesses and other markets'.

The deal made Three Gorges EDP's biggest shareholder, and the Chinese company also arranged a €2bn (US$2.8bn) credit facility provided by a Chinese financial institution for up to 20 years. 'This liquidity is very important, as everybody can imagine, and we believe that this deal has a very positive impact on EDP's credit profile', said Mexia (Mexia 2011).

In August 2013, Japanese trading and finance house Marubeni agreed to take a 25 per cent stake in pioneering renewables developer Mainstream

Renewable Power for a reported €100m (US$138m), in a deal that will also see Marubeni take equity stakes in Mainstream's projects. Officials from Marbueni said that they were particularly interested in Mainstream's widespread participation in UK offshore wind projects.

Mainstream founder and CEO Eddie O'Connor said the deal means that 'We can say to a country or a customer, "Not only do we have the capability to develop the project, but we can also bring along someone who may invest in the equity and ownership"' (Stromsta 2013b).

The deal came on the same day that Marubeni agreed to buy a 50 per cent share in the 3.3GW of Portuguese generation assets owned by GDF Suez – including wind and solar PV. The company had already purchased a 49.19 per cent stake in DONG Energy's 172MW Gunfleet Sands offshore wind project, as well as stakes in onshore wind projects such as EDF's 205MW Lakefield wind farm in Minnesota.

'Marubeni needed early-stage development resources on a global basis', Marubeni Europower president Hiroshi Tachigami told *Recharge* from London. 'One of the things we see in Mainstream is the "offshore story", where Marubeni seeks to locate itself in all stages of the offshore wind value chain … from feasibility studies, pre-construction, construction, operation and decommissioning' (Publicover 2013a).

As European utilities and developers find their ability to finance big project portfolios increasingly under pressure, we can expect to see more deals with capital-rich Asian companies.

Meanwhile, increased competition among turbine manufacturers wasn't just coming from established Western engineering giants. The wind industry was about to experience the China effect.

3   China shakes the wind industry

Across the arid plains of Gansu, an endless array of wind turbines stretches towards the horizon. The turbines are part of a series of 'large wind power bases' that are growing across northern China, from Xinjiang in the West to Jilin in the East. The onshore wind bases had a combined capacity of 39GW (yes, that is gigawatts) at the end of 2012, and are set to reach 165GW by 2020. The scale of China's ambition is, as ever, breathtaking.

Annual installations in 2002 were 66.3MW, but they had risen to 1.29GW by 2006. Installations mushroomed to 13.8GW in 2009, 18.93GW in 2010 and 17.6GW in 2011. China overtook the US for the first time as the world's largest market by cumulative installed capacity in 2010, and by the end of 2012 had a cumulative capacity of 75,324GW, compared to 60,007GW in the US.

Foreign turbine manufacturers, led by Vestas and Gamesa, had taken a lead in starting China's wind-turbine industry in the hope of gaining huge new markets for their products. Vestas in particular made important contributions to China's industry through activities like carrying out studies – at its own expense – of China's grid requirements. In the middle of the last decade, Western turbine suppliers including Vestas, Gamesa, Suzlon and GE had carved out more than 70 per cent of the market.

In a scenario familiar to other industries, however, China's wind boom was accompanied by a rapid expansion in domestic wind suppliers. By 2010, the top Chinese companies accounted for around 90 per cent of the local market and were challenging established suppliers in the ranking of the top ten global turbine suppliers. In 2011, the then biggest Chinese supplier, Sinovel, which we will look at in detail in a moment, was challenging Vestas for the number one spot.

How did they do it? On the one hand, Chinese companies learnt fast. Often taking licensed Western designs for a base, they were able to build up large-scale manufacturing and execute large-scale projects at breakneck speed. And they were able to supply turbines at a price level that put severe pressure on the margins of the established Western players. Analysts estimate that prices have fallen by more than 40 per cent since 2008, as companies have expanded and the industry has attracted new entrants prepared to offer low prices in return for initial sales.

They were also helped by a series of rules that were aimed at favouring local content during the early years of the industry and, as we shall see, were able to leverage central and regional government contacts on an ongoing basis. While Western companies were able to continue to win smaller contracts from utilities, the big wind-based projects were effectively reserved for Chinese companies only, allowing a number of local companies to achieve massive sales. Finally, Chinese turbine manufacturers and utilities were able to take advantage of almost unlimited funding from state-owned financial institutions, such as the China Development Bank, to set up factories and build projects.

The massive expansion of turbine supply implied a big compromise in terms of quality, the effects of which have yet to be fully played out. China saw a large number of industrial accidents involving turbines during 2009–12, including tower collapses, blades shearing off, fires and electrical accidents that saw a number of workers killed.

As the Chinese Wind Energy Association describes:

> In recent years, the disadvantages related to high-speed development began to reveal themselves. Starting from the second half of 2009, a series of incidents happened in succession, including tower collapses, blade ruptures, nacelle fires, personal electric shocks and engineering accidents, leading to more than ten casualties and damage to more than ten sets of generator equipment.
>
> (Li *et al.* 2012: 63)

Mechanical failure was so rife at some of the big northern wind farms that operators sometimes had as many as one person per turbine working on operations and maintenance (compared to a couple of technicians for a large wind farm in the West). Stories abounded of technicians sleeping in the nacelles of machines to stave off cold weather conditions.

What allowed such widespread quality problems to exist was the fact that large utilities were more concerned about reaching almost recklessly ambitious internal and government targets as quickly as possible, and at as low a cost per MW installed as possible, rather than considering the cost of energy per MWh over the projects' full life cycles. Indeed, it is debatable if many of the projects that were installed in the boom years will really have the 20-year life cycles that are typically the base for calculating wind farms' returns and values.

China's existing power system was also finding it hard to digest wind power, partly due to the huge new volumes and partly due to the low level 'grid friendliness' of Chinese turbines. Since 2011, an increasing number of incidents of turbines tripping off the grid have occurred, with 193 incidents of this kind across the whole country during January to August 2011 alone, according to the CWEA. Analysis carried out by the State Electricity Regulatory Commission identified four major issues: a) most wind turbines have no low-voltage ride-through capability; b) there are many quality problems in wind farm construction; c) connection of large-scale wind farms could threaten the stability and safety of the power grid; and d) wind farm operation management is weak.

By 2011, China's wind industry was facing massive growing pains. The country's grid system was buckling under the strain of bringing on the new wind capacity, particularly because the bulk of projects had been located in the arid north and west, far from the densely populated – and transmission system-dense – demand centres in the eastern and southern coastal areas. According to the China Electricity Council, 47.84GW of capacity was connected to the grid, out of a total installed capacity of 66.4GW, while 10 billion kWh of wind power was not produced due to curtailment. Regionally, the amount of wind electricity curtailed in the north and northeast were the largest, accounting for 57.20 per cent and 38.33 per cent of the total wind electricity curtailed nationwide, respectively.

In the latter part of 2011, the market began to slow down fast. It soon became clear that a massive build in inventories was taking place, with OEMs struggling to place their production. It also became clear that China's wind boom and easy financing had attracted too many turbine manufacturers into the market. According to analysts, there were over 80 Chinese OEMs in 2010 and around 70 in 2011.

According to BTM Consult, manufacturers collectively had around 28GW of capacity in 2011, compared to annual installations of 17.6GW in 2011 and 12.96GW in 2012. Although the level of installations was still the second highest in the world – as we have seen the US bounced back into first place with a record 13.2GW – 2012 became 'the year of adjustment' for China's wind industry. Many smaller turbine manufacturers simply stopped producing and shifted their activities back into other industrial areas, while, as we shall see, the adjustment had differing effects on the fortunes of companies within the top 15 companies, which accounted for 93.9 per cent of China's total installations in 2011.

The government rushed to build a series of major long-distance transmission lines, connecting the northern areas to the demand centres, while encouraging growth closer to demand centres, and slowing down the development of the big northern wind bases. As the Chinese Wind Energy Association says:

> On the one hand, the government was looking into different options to increase the consumption of the wind in [the producing] regions, as well as to increase the transmissions to the neighbouring regions with higher electricity demand. On the other hand, the government also realised that before the problem of electricity consumption or transmission can be solved overnight, it would be good to slow down the process of Wind

Bases while starting to develop wind in the central and east area, where wind resources may not be prominent but electricity load is higher and transmission infrastructure is robust.

(Li, J. *et al.* 2012: 13)

This sent the surviving wind-turbine manufacturers scrambling to adapt to a new type of market. First, developers were now in more of a position to pick and choose when selecting turbines, but new government regulations and increasing sophistication among operators meant in many cases a renewed emphasis on quality and cost of energy over project life-times.

Second, developers shifted their investment towards projects to transmission-rich areas closer to demand centres, which meant moving to sites that on average were less windy. The turbine manufacturers that were in a position to do so began to produce models with larger rotor sizes designed to capture more wind at lower speeds.

Third, manufacturers began to step up efforts to place production in international markets, as part of an ongoing 'go global' strategy encouraged by the government. Chinese companies like Sinovel had already become a fixture at international trade shows with impressive pavilions, and companies attempted to close numerous deals, from Africa to Eastern Europe to the Americas. There was continued – although unfounded – speculation through 2011 and much of 2012 that a Chinese company could be preparing a major acquisition of a European wind-turbine producer, even on the scale of Vestas, with Mingyang among the names mentioned.

It is interesting to look at the rollercoaster ride of the Chinese wind-turbine industry through the stories of two companies; Sinovel, which was the fastest-growing wind-turbine company in the world during the boom years and briefly challenged Vestas for global number one spot; and Goldwind, which is the current number one in China.

Sinovel's rise and stall

The meteoric rise of Sinovel was one of the most visible products of the boom in the Chinese wind industry.

In 2011, just five years after the company was set up, it had become the world's second-largest turbine manufacturer – with 4.4GW of annual installations and 11 per cent of the global market – and was threatening to overtake the leader, Vestas. Its initial public offering (IPO) that year was the most expensive ever on the Shanghai stock exchange, with shares being listed at 90 yuan (US$14) each.

By 2013, however Sinovel's market share had plummeted, its shares had fallen to less than five yuan; it was being sued by former components supplier AMSC for US$1.2bn and being investigated by Chinese authorities for suspected violations of securities laws, in a case that reportedly involved inflating its revenue numbers (see Lee and Publicover 2014).

At the time of writing, the company had reported it was expecting its 2013 net loss to balloon to 3bn yuan (US$495.4m), marking its second straight year of red ink. The loss was due to a decline in total installations, delayed orders and postponed payments, it said in a statement to the Shanghai Stock Exchange (SSE). The slump in Sinovel's fortunes marked a precipitous reversal of fortunes for a company that had been close to reaching the pinnacle of the global wind-turbine industry.

Sinovel was founded in 2006, and started life as an electrical-equipment subsidiary of state-owned industrial group Dalian Heavy. With no prior experience of manufacturing wind turbines, Sinovel licensed technology to produce a 1.5MW turbine from German turbine manufacturer, Fuhrländer. Less than two years later it had been awarded a huge 1.8GW order for the first phase of the Jiuquan wind park. 'To go from producing 50 units one year, to 1,000 units two years later was impressive', an industry source told *Recharge*. 'A European firm would never have expanded this fast.'

The key to Sinovel's growth was a perfectly timed product offering – no other Chinese company had a 1MW+ size turbine on the market – while parent company Dalian Heavy had plenty of factory capacity and engineering expertise.

Sinovel's then Chairman and President, Han Junliang, was extremely effective at marketing the company to investors, customers and policymakers. Junliang had an impressive array of high-level political connections, and his unprecedented efforts to nurture these relationships left others floundering in Sinovel's wake.

For instance, a subsidiary of New Horizon Capital – co-founded by Wen Yunsong, the son of Chinese Premier Wen Jiabao – bought a 12 per cent stake in the company. To shore up his government connections, Han reportedly lavished gifts on officials, typically flying expensive fresh seafood from Dalian to Beijing for the Chinese New Year, or handing out US$1,000 bottles of whisky on other occasions.

Sinovel enjoyed a strong connection with Zhang Guobao, the former vice-chairman of the National Development and Reform Commission (NDRC), the country's powerful policymaking body. In 2004, Zhang Guobao was tasked with overseeing the restructuring of heavy-industry groups in northern China. 'He had a mission to support all heavy industry to do something new, to re-establish growth', says a source who works for a rival turbine maker. 'This [Sinovel's founding] was certainly a project in line with government policy.'

Zhang was a strong supporter of 'national champions' and everything suggests that Han knew he would win government support for Sinovel's larger turbine. The state-owned power industry is 'sensitive to the country's development', a former Sinovel employee explained to *Recharge*. 'If a local producer can make bigger turbines, they will buy them. They want to support local production and it makes the government look good to have big turbines.'

UNITED NATIONS CLIMATE CHANGE CONF

Figure 3.1 Sinovel's Chairman and President, Han Junliang (Christoph Bangert/ Danish Wind Industry Association)

In 2008, Zhang became director of the newly created National Energy Administration, just as China launched its first round of concessions for the mega-wind bases such as Jiuquan. The government did not specify a requirement for Chinese turbines, but no foreign firms won significant orders from these tenders.

Of the local players, Sinovel took the lion's share. After coming away with 1.8GW of orders in the 2008 tender – close to half the total – it followed up with 2GW from the Inner Mongolia and Hebei concessions in 2009, and a further 1.35GW the following year.

These concessions played a large part in creating Sinovel's huge order backlog, and fuelled Han's ambitions for further growth. By the end of 2010, Sinovel had orders in hand of more than 14GW. Han told investors ahead of the company's IPO that Sinovel would see a compound annual growth rate of more than 30 per cent for the next five years. Indeed, its sales had surged by 48 per cent during 2010.

But, around the same time as Zhang's retirement in early 2011, the 'year of adjustment' hit, and Sinovel's management realised it had become too dependent on large government tenders and a handful of key customers.

Many of China's large concessions were taking longer than expected to be built, and Sinovel's inventories were piling up. Realising it had vastly overestimated its expected output, in March 2011 Sinovel abruptly refused component shipments worth US$70m from its key converter supplier AMSC. This dealt a body blow to AMSC, which at the time of writing was suing Sinovel in Chinese courts for these unpaid deliveries, as well as a further US$700m for unfulfilled contracts. Worse, Sinovel's involvement with AMSC, which was co-designing turbines with the company as well as supplying components, also led to an industrial espionage case that would rock the international wind industry, and cripple the Chinese company's plans to become a global player

According to allegations from AMSC, which has three civil suits in Beijing, seeking a total of US$456m in compensation for intellectual-property theft, Sinovel stole software codes needed to upgrade its control systems to meet new grid standards in China. In September 2011, an Austrian court sentenced former AMSC Windtec employee Dejan Karabrasevic to a year in jail for 'fraudulent misuse of data' after concluding that he had passed the company's software codes to Sinovel in exchange for €15,000 (US$19,500). The evidence suggests that Sinovel was preparing to replace key components supplied by AMSC with locally manufactured equivalents, and needed to crack the software codes in order to do so.

Global media coverage of the lawsuits has hit Sinovel's international reputation, scuppering its largest overseas deal to date – a 1GW contract with Ireland's Mainstream Renewable Power – and destroying any chance of making sales in the US.

'We are all bound by law in Europe, and I suspect that I could be put in prison for receiving stolen goods', Mainstream founder and CEO Eddie O'Connor told reporters in Amsterdam in November 2011, shortly after cancelling the deal. He added that he had been doing everything he could to persuade Sinovel to take part in arbitration, and that Mainstream would be keen to work with the Chinese company in the future.

In Brazil, the threat of legal action from AMSC led to developer Desenvix filing a court order against Sinovel, demanding the right to inspect the 23 turbines it had ordered, to ensure they did not contain intellectual property 'stolen' from AMSC. (Desenvix dropped the case after it emerged that the turbines used control systems from US firm Emerson Electric.)

The Mainstream loss came as the firm's domestic orders were also shrinking. Large concessions had ground to a halt following a series of major accidents at wind farms in Gansu and Hebei that raised concerns about the safety of Chinese turbines. 'Sinovel benefited a lot from the large wind bases, but they tended to get larger volume orders from fewer customers', said Justin Wu, a wind analyst at Bloomberg New Energy Finance. 'This means they have been more affected by the slowdown in the north. And they had exposure to fewer clients than Goldwind, for example' (cited in Patton 2013).

The toll from a torrid 2011 was revealed in Sinovel's full-year report. Sales had dropped by almost 50 per cent, while profits were down 73 per cent to

776m yuan (US$125m). The numbers significantly worsened in 2012, with a loss of 582.7m yuan (US$94.5m). The company was also losing market share rapidly, falling to third in the China rankings – behind Goldwind and United Power – in 2012 and much further down the rankings in 2013 with a China market share of less than 6 per cent.

Sinovel's Vice-President Tao Gang described the market environment for the company from 2012 onwards as 'very tough'. Referring to falling turbine prices and the prospect of delayed payments from customers, he said: 'On one side is the volume, and on the other side is your profitability. The more you install, potentially the more money you lose, either currently or in the future' (cited in Patton 2013).

One symptom of the change in Sinovel's position it that orders from developer Huaneng, its largest customer, almost entirely ceased. In 2010, Huaneng bought about 40 per cent of Sinovel's turbines, according to data compiled by Beijing-based consultancy Azure International. In 2011, it consumed only 28 per cent of Sinovel's reduced delivery volume, and orders fell even further in 2012 and 2013. 'In the beginning, there wasn't much choice of suppliers for 1.5MW turbines so we bought a lot of Sinovel turbines', Jiang Xueyong, investor relations manager at Huaneng Renewables, told *Recharge* (*Recharge Magazine*, 6 January 2013).

But Huaneng placed few orders with Sinovel in 2012. 'Last year, we were building most of our projects in the south on high plateaus, so we were looking for turbines suited to those conditions. We've tended to buy a lot of turbines from CSR [China South Railway's wind unit].' In a widely reported speech at the 2011 China Offshore Wind conference in Shanghai, Xie Changjun, the president of China's leading developer Longyuan, attacked Sinovel for its poor quality record.

The question of quality and safety has also negatively affected Sinovel's sales. In a report to investors in September 2011, Daiwa Securities analysts in Hong Kong concluded that about 40 per cent of the 27 wind-turbine accidents they counted during 2010 had involved either Dongfang or Sinovel models. More accidents occurred in 2011, including electrical accidents that killed workers, as well as a disastrous collapse of a crane at Sinovel's Jiuquan factory that killed a high-ranking local communist official.

In March 2013, Sinovel brought in a former director of the Shanghai stock exchange and long-time Sinovel shareholder, Wei Wenyuan, to replace Han Junliang. Wei started by closing three facilities – the Dalian factory, a sales and service centre in Jiangsu Nantong and a factory under construction in southern Guizhou province. He also put 400 staff on 'paid holiday' – mostly from the research and development (R&D) department. Two months later, though, Wei was out.

Han is said to have also stressed that he is 'still in control', even though he was voted out of his role as president at an August 2012 board meeting, and remains chairman at the time of writing. There has been industry speculation, though, that Han – a one-time billionaire thanks to his

12 per cent stake in the company – could cash in his shares and head for new pastures.

To add to Sinovel's terrible 2013, Chinese authorities initiated an investigation into the company in May, reportedly for inflating its revenues in its stock-market filings. The probe was still under way at the time of writing, and had reportedly turned into a criminal investigation.

Sinovel appears to have significantly cut back its ambitions, such as building a 10MW turbine and becoming the world's leading turbine manufacturer within five years of its stock-exchange listing. It also seems to be set to fall further down the rankings of Chinese turbine manufacturers by market share, as companies like United Power, Ming Yang, SeWind and XEMC continue to challenge, but it still has significant strengths.

Sinovel has a leading position in China's nascent – but fast-growing – offshore market. At a conference in Beijing in October 2013, Vice-President Chen Danghui said 'We've received lots of negative coverage, but we can overcome (these difficulties) because our core people are still around.' The company is working hard to fill the gap left by AMSC, working with Danish control electronics provider Mita-Teknik to cover the gap left by AMSC, and third-party testing agencies GL and TÜV Nord to certify its turbines, a key step in ensuring credibility in international markets.

> Sinovel's Senior Vice-President Tao Gang says 'We have rich experience offshore', and points to the connection of its second 6MW turbine to the grid at a site run by developer Luneng in Zhangbei county, in northern China's Hebei province, as evidence of its technology lead in offshore. Sinovel plans to deliver 17 units to a 102MW pilot offshore project in Shanghai's Nanhui district in the first quarter of 2014, as part of a 1.76bn yuan (US$288.5m) project scheduled for completion by September 2014. Tao points out that Sinovel was the first Chinese manufacturer to roll out 3MW, 5MW and 6MW turbines.
>
> (Cited in Publicover 2013b)

With 2013 expected to show further big financial losses (net losses were 699m yuan (US$113m) from January to September compared to a loss of 269m yuan (US$44m) during the same period in 2012), and the legal dispute with AMSC unresolved at the time of writing, Sinovel faces a hard slog to get out of its current problems.

Goldwind plays the long game

In contrast to Sinovel, Xinjiang-based Goldwind Science and Technology has managed to maintain its market share and avoid major corporate upheaval and legal controversy. While its growth was not as spectacular as Sinovel's, Goldwind enjoys some notable advantages, in particular because it has significant design resources – and owns its own intellectual property rights – through its acquisition of German turbine manufacturer and designer, Vensys.

Wu Gang, an early pioneer of Chinese wind-power technology who worked on some of the first demonstration wind farms, was instrumental in setting up Goldwind, the country's first wind-turbine company, in 1997. By 2006, Goldwind controlled around 33 per cent of China's total market. It became a listed company in 2007 in a US$216m IPO on the Shenzhen exchange, and this put it in a position to acquire a 70 per cent stake in Vensys in 2008. It carried out a further IPO on the Hong Kong exchange in 2010m, raising US$917m.

Goldwind had begun a transition from gearbox-based technology to a permanent-magnet direct-drive (PMDD)-based design in 2003, when it partnered with Vensys to co-develop the GW1.2MW model, which subsequently evolved into the 1.5MW model. Since the acquisition, Goldwind has developed 2.5MW, 3MW and 6MW machines.

As well as designing Goldwind's machines, Vensys has licensed successful direct-drive designs to companies such as Impsa in Brazil and Regen Powertec in India. Goldwind has also been unique among Chinese manufacturers in carrying out a steady expansion into the 'mature' US market, as well as Australia and other countries.

Wu Gang – now Chairman – often emphasises the importance of a truly international technological community around wind, from his earliest visits to the Riso wind-research centre in Denmark. He told me in Germany in late 2012, 'We have an open mind and we try and learn from different countries' cultures, so that is quite different from other Chinese companies' (personal communication).

Wu says that he sees a 'global' approach as key to his company's strategy, and says it encompasses five elements: internationalising the company's technology and products; internationalising its human resources; expanding into international markets; gaining access to international capital; and adopting an international management system.

A key area for Goldwind's international expansion is the US. In contrast to other Chinese turbine manufacturers that have announced big projects in the country which have then failed to materialise, Goldwind's emphasis has been to construct – and help finance – mid-sized projects such as the 109.5MW Shady Oaks project in Illinois in order to establish a track record, so that developers and utilities can see the effectiveness of its turbines. 'We have good communication with local customers, and we can help on the investment side of the project so they can easily see the performance', says Wu.

Goldwind is also aggressively seeking to expand in Australia, where it has set up a Sydney office; in Latin America, where it has secured projects in Chile and Panama; and in Pakistan. It is also leveraging the relationship between major shareholder Three Gorges and the Portuguese wind-development giant EDPR to jointly take part in tenders with the latter.

Goldwind's plan had been to dedicate around one-third of its production to international markets by the end of 2015, with major growth expected to come in other Asian countries, Africa, North and South America, and Australia. However, intense competition in global markets means that in common with other Chinese OEMs 'Go global' is going slower than officials had expected. But plans for a US factory are on hold, and Wu says the company has not taken any firm decisions about setting up assembly operations outside China, although he states proudly that Goldwind US uses more local components than other 'local' manufacturers.

At present, the focus is on getting through the 'adjustment' years. Goldwind was quite quick to see that the Chinese industry was about to go through major changes, and began to change its plans in 2011 to adapt to a period of intense competition and a reduction in the pace of expansion. 'We know it will be a very cold winter for the wind industry so, since last year, we have been preparing how to spend the wintertime', Wu told me in Husum, Germany, in late 2012. 'Sometimes winter can be good for companies to improve the quality of their management.'

In contrast to some of its Chinese competitors, Goldwind has put a continuous effort into R&D, ensuring that its technology keeps up with the latest international level. At the same time, it has strictly controlled employee numbers at around 4,000 employees, making it a very lean company for its size in terms of MW delivered. In 2012, it also outsourced its blade-manufacturing activities, selling its Tianhe subsidiary while maintaining a stake.

The company has also become more 'vertically integrated' by expanding its activities down into project development, financing and engineering procurement and construction (EPC) services – all key activities in a period of market slow-down – which make the company's turbines more attractive to potential buyers.

Similarly, while Wu is confident of the tremendous strength that the Chinese wind market can provide to his company when faster growth resumes in 2014/15, pointing to the strong underlying support from government policymakers, the company has also had access to large-scale loans from the China Development Bank to finance its expansion.

However, Wu says Goldwind is not in the market for acquisitions, despite bouts of speculation. 'We are focused on new technology', he says. 'We don't want to buy companies with traditional technology. We are looking at the future.'

Goldwind's roots in the remote northwest region of Xinjiang have taught it to keep an open mind, seek co-operation and expand cautiously, he says. 'Our approach is to grow step by step. We are different from some companies that see everything as simple', Wu says.

Where now for the Chinese manufacturers?

After the two years of adjustment in 2012 and 2013, the Chinese wind-turbine market is expected to resume a faster pace of growth in 2014, and should soon be reaching a steady level of installations of a whopping 20GW per year. This does not mean, however, that things will become easier over night for turbine manufacturers, given the intense competition between a leading group of 15 companies, and continued overcapacity in the market. The last couple of years have seen some dramatic events; including the precipitous fall down the rankings of Sinovel and of Dongfang Electric.

As we have seen, Goldwind has been more successful at maintaining market share and corporate stability in tough times, and is the undisputed number one in China by market share.

But although we have talked quite a bit about quality and intellectual property rights, the main factors in the fortunes of Chinese turbine manufacturers have been relationships with the 'big five' wind developers.

One notable trend has been the growth of state-owned turbine manufacturers with close links to utilities. The big winner in the turbine market, aside from Goldwind, has been Guodian United Power, which became the second-largest Chinese OEM in 2012 and achieved a market share of almost 10 per cent in 2013. Guodian of course is also the owner of Longyuan, China's largest developer. The company has big plans, and has developed a number of new turbine models including a 6MW offshore machine.

Rail company CSR and shipbuilder CSIC are two other big state-owned companies that have prospered and risen in the rankings, while other state-owned power companies such as Datang and State Grid have bought or set up their own turbine companies.

Analysts say that the growth of these companies and the existence of a series of other state-owned turbine manufacturers could delay the consolidation of the industry that is led by a small group of leading companies. Despite clear overcapacity, the state-owned companies continue to view the wind industry as an opportunity for diversifying. 'State-owned companies are entering into sectors categorised as the strategic emerging industries, especially in the economic downturn', Reggie Lai, deputy managing director at consultancy APCO, told *Recharge* (Patton 2012). These companies regard wind power as a sector with long-term government support, adds Lai. 'There is a policy basis to invest in this sector.'

As well as the state-owned players, several other companies have taken advantage of the relative decline of Sinovel and Dongfang. These include NYSE listed Ming Yang, which has grown aggressively and is developing a series of innovative turbines with German designer, Aerodyn; the highly regarded Envision, XEMC and Sewind (the joint venture which Shanghai Electric formed with Siemens in late 2011).

How long consolidation will take to happen in these circumstances, and who will be the ultimate winner, is hard to predict. In broad terms, one could predict that the companies that are able to develop in terms of sophistication of products and services – the area of providing operations and maintenance services for China's growing fleet of turbines – and being able to extend turbine life cycles will be in a position to gain market share and grow in the coming period.

To some degree, the outcome of the battle will depend on turbine quality and good business strategy and management. But it will also depend to a great degree on key relationships with the big developers, as well as government industrial policy. Policy committees have been proactive in forcing through the consolidation of other industries in China – for example telecoms – and there is no reason to expect that they will not be active in forcing through consolidation of the turbine industry. Whether this takes the form of favouring a certain number of national champions, encouraging mergers and acquisitions or simply pushing through market changes that will privilege real competition in terms of cost of energy – or all of the above – will be the moot question of the next couple of years.

The foreign OEMs try to hang on

Meanwhile, what is the situation for foreign manufacturers in China, and what are their chances of survival?

As we saw, Western turbine makers were quick to see the huge potential of the Chinese wind sector, and had carved out more than 70 per cent of the market by the middle of the last decade. By 2012, their market share had been eroded to only 9.8 per cent, according to BTM Consult. The top four non-Chinese manufacturers in China together had installed about 1GW in 2013, out of a total market of around 13GW, according to Justin Wu, lead wind-energy analyst at Bloomberg New Energy Finance in Hong Kong.

The reasons for the decline are myriad. Patrick Dai, a Hong Kong-based analyst at Macquarie Securities, puts the rise of Chinese OEMs down to 'a combination of competitive pricing and advanced technology, flexible customisation, easy servicing and relationships with the local governments'. Competition has been intense, particularly given overcapacity among the Chinese players, and the hiatus in the market during 2011–12 has led to intense pressure on margins.

In the past few years, a number of Western companies have decided that the Chinese market is not for them.

In October 2009, Spain's Acciona sold its 45 per cent stake in its joint venture with a subsidiary of state-owned China Aerospace Science and Technology Corporation (CASC) – just three years after establishing a plant in Nantong, which it claimed at the time was China's largest turbine plant.

In 2012, Indian-owned Suzlon – once a top-ten player in the market – announced that it was selling its manufacturing unit in China, despite top officials claiming just a year earlier that it was able to compete with local companies on costs and pricing. The sale of a 75 per cent stake in Suzlon Energy Tianjin to Poly LongMa Energy (Dalian) was eventually completed in September 2013, after an earlier deal with China Power New Energy, announced the previous June, fell through. Suzlon subsidiary, REpower, announced it was leaving the Chinese market in September 2011 because of 'increasing trade protectionism', and closed its majority-owned factory in Baotou, Inner Mongolia.

In July 2013, US manufacturer GE ended the joint venture (JV) it had formed with Chinese power-equipment group Harbin Electrical Machinery in September 2010, although it says it remains committed to the market.

Meanwhile, others have discreetly shelved plans to expand in China. These include Nordex – which tried unsuccessfully to conclude a joint venture with local developer Huadian in 2012, and subsequently scaled down its local operations – and Alstom Wind, which had planned to set up a new Chinese plant.

Those remaining say they are determined to stay, arguing that as the Chinese market matures – in terms of customer needs, geographical diversification and pricing, with widespread consolidation expected among local players – opportunities will begin to grow again. 'Once the transmission bottleneck is solved the market will take off again', says BTM Research Director Feng Zhao, who adds that the level of annual installations could reach 20GW by 2017. 'So if the Western companies can just maintain their market share where it is, it's not so bad', he says. 'If you don't have the patience and you are not there, you are losing out on the business' (Backwell and Publicover 2013).

For Zhao, the key to success in China is offering the latest technology, particularly as the market is becoming more sophisticated and projects shift south to sites with lower wind speeds. 'I think we can see across emerging markets that if you arrive late and with old technology it's not going to work', he says (ibid.).

The other key factor is relationships with the major developers. 'If you look at the data, you can see that Vestas and Gamesa both have at least two relationships each with major customers', he points out (ibid.). The remaining foreign OEMs – Vestas, Gamesa, GE and Siemens – have widely different strategies, while there is a question mark over GE's future.

Vestas

Vestas was an early front-runner in China, and had installed more than 4GW in the country by the end of June 2013. Its ranking in China fell from sixth place in 2010 to tenth in 2012, when it accounted for just 3.2 per cent of the market. But it has strong relationships with developers such as Datang, China Resource New Energy and CGN Windpower. And it has secured repeat business from many of the companies it has supplied.

Its attempt to sell locally produced V52-850kW and V60-850kW machines was a failure, and it closed its Hohhot factory, leaving the 'kilowatt' platform in summer 2012. But Vestas has maintained its Tianjin factory, as well as its Beijing and Shanghai offices. And it is winning a steady flow of orders for its 2MW V90 and V100 turbines, including recent orders from Chinese gas group Hanas New Energy and Hebei Construction and Investment Group.

Vestas' turbines feature yaw backup systems to protect blades from the typhoons that frequently lash China's southeast coastline, where it has been a particularly strong player. It has long won projects in Fujian province, for example, claiming roughly 600MW of the province's 1.6GW installed capacity by late 2012. And it should be well positioned to win orders when offshore takes off along the eastern seaboard, as Fujian, in particular, wants to install 500MW offshore by 2015.

The challenge companies like Vestas face is capturing niche segments such as offshore, low-wind and high-altitude projects with their technologically advanced turbines, says Wu. 'When you're able to source from China's supply chain itself, you're not necessarily offering those products', he says. 'The question for foreign manufacturers, really, is whether or not they can introduce their new products into China … you're not competitive on cost when you move into those turbine segments' (Wu, quoted in Backwell and Publicover 2013).

In recent years, Vestas has also shown that it is prepared to scale down operations and stop participating in tenders if prices are not attractive, as it openly stated in 2012. The level of participation in China is subordinate to the wider aim of returning the company to profitability.

Gamesa

Gamesa is widely tipped as one of the foreign OEMs most likely to continue to do well in China, mainly due to its strong relationships with some of the country's leading utility players, such as Longyuan and Huadian. It was ranked in ninth place in 2012 with 493MW, according to BTM Consult figures. '[We] can provide experience in niche markets and in specific technologies beyond what the local manufacturers can provide', says Gamesa China's boss José Antonio Miranda (Backwell and Publicover 2013), pointing to issues such as resistance to turbulence and extreme weather such as typhoons, and developing high-altitude sites. Gamesa is also looking at providing O&M services, offering improvements in performance and the extension of turbine life cycles.

One key aspect of Gamesa's strategy is its approach to introducing technology into China, with an approach to intellectual-property (IP) protection that could be described as less obsessive than some of its peers. 'Two years ago, we decided to launch our products all at once in all regions', says Miranda. 'It's clear to us that to compete we have to come with our most up-to-date technology' (ibid.).

From the point of view of IP protection, Miranda describes China as 'no different'. 'In any market, your new product can be an inspiration, but we have not seen any IP theft here' (ibid.). Miranda notes that Gamesa has been producing in China since 1995, has a 'very developed' local supply chain and produces large turbine parts such as hubs that are sent to other parts of Asia and Europe.

Gamesa is also eyeing the offshore market in the medium term, and says it would look to deploy its new 5MW turbine from 2016 onwards.

In the future, Miranda estimates that there will be no more than 'two or three' foreign manufacturers operating in the market, but that there are 'no special difficulties' for non-Chinese players. 'The domestic companies will improve their quality and that has a cost, and the international companies will become more specialised in niche areas, and there won't be such big differences in domestic and international markets', he says (ibid.).

Siemens

Siemens' strategy differs from Vestas' and Gamesa's in that the company has long said that it would focus on a JV with a local heavyweight industrial partner.

In late 2011, Siemens' plans came to fruition when it announced an agreement with long-term partner, Shanghai Electric. The two companies agreed to jointly invest €169.1m (US$226m) to set up two wind-power equipment joint ventures in China – Siemens Wind Turbines (Shanghai) and Shanghai Electric Wind Energy (Sewind) – with blade and nacelle plants in Shanghai producing Siemens' SWT-2.5–108 model.

The JV is producing turbines for onshore projects, including the 50MW Guangrao wind farm, 400km south of Beijing in Shandong province. An official from Shanghai Electric said in February 2013, when the project was completed, that the venture plans to install about 300MW of onshore wind in China this year. It is not clear if it is on track to meet this target.

Unsurprisingly, given Siemens' dominance in the sector, Sewind says that it hopes to claim around 25 per cent of China's offshore market by 2015.

Siemens supplied 21 of its 2.3MW turbines for Longyuan's Rudong inter-tidal wind project, which went operational in May 2012. Shanghai Electric had also won a tender in 2011 to supply 26 of its 3.6MW turbines to the 101MW second phase of the Donghai Bridge wind farm off Shanghai, and will supply the 200MW Luneng Jiangsu Dongtai project. Sewind says it expects to sell 75 of its SWT 4.0–130 turbines in 2014, and double that amount in 2015.

The logic of Siemens' move is clear, but there is some scepticism about how successful the JV will be in practice. 'The question is, how many joint ventures have worked?' says Feng, pointing to the break-ups of GE and Harbin, Acciona and CASC, and others. 'Basically, it's a battle between technology and market', he adds. 'The Chinese want cutting-edge technology, but for foreign companies, the IP is the value and they want to hold on to it. It's a tough bargaining process' (Backwell and Publicover 2013).

GE

There is some uncertainty about the strategy of US turbine giant GE, after it ended its joint venture with Harbin Electric in July 2013, less than three years after it was set up.

In 2012, GE ranked 17th in China, with 90MW installed, according to the Chinese Wind Energy Association, compared to 11th place with 408.5MW of deliveries a year earlier. According to Patrick Dai, GE does have a strong track record with east-coast projects, and the company was one of Longyuan's top five suppliers in 2012. However, Dai adds, 'I don't see any progress' (Backwell and Publicover 2013).

Wu believes GE's 'reasonably localised supply chain' (ibid.) could help it sustain its local operations. 'They're still quite active in China', he says (ibid.), noting that it is still winning new contracts and buying key components such as gearboxes from local suppliers, including China High Speed Transmission, one of the country's largest makers of wind-turbine gearboxes.

It also has the products and technologies to compete for low-wind and high-altitude projects in China. 'Certainly, GE's class-three wind-speed products have been quite successful in the West', says Wu (ibid.).

'Go Global' – postponed not cancelled

The idea that Chinese wind-turbine companies were about to move into international markets and drive incumbent manufacturers out of the market was almost an obsession among many observers of the wind industry during 2010–12. Financial analysts trying to take a mid-term view of wind stocks, in particular, were drawn by the idea, particularly as Chinese companies were effectively wiping out incumbent European and US companies in the solar PV sector at the time.

However, by 2013 it was clear that the predicted wave of Chinese turbines flooding markets and widely predicted large-scale acquisitions by Chinese companies had not materialised. 'The go-global strategy has failed', says HIS-EER Research Director Eduard Sala de Vedruna.

The explanations fall into two categories. On the one hand, go-global has not happened for turbine manufacturers because their largest clients – big Chinese developers like Longyuan and Datang – have been slower than expected in carrying out their go-global plans, as they discover the difficulties of operating in a foreign environment and often see relatively higher returns from local projects.

On the other hand, it was clear that most of the Chinese turbine companies were not adequately prepared for operating internationally. The most serious issues were those of lack of a track record and the dubious quality reputation of some Chinese machines, coupled with the fact that many of these had not been certified by international companies like DNV or Garrad Hassan. In 2011, a time of intense Chinese activity in the US, developers complained that some Chinese turbine manufacturers did not have full documentation in English or wanted provisions for dispute resolution to be based on legal jurisdictions. 'Culture' was also a problem. Leading turbine companies like Sinovel often had enormous stands at international trade shows, but had staff who were unwilling or unable to engage with prospective customers when they approached.

There were some high-profile failures, such as a project to build a big US wind farm and manufacturing plant – with high-profile political support – by Chinese company A-Power. But it has been the Sinovel case that has shown what can really go wrong, with the AMSC dispute killing the biggest Chinese turbine deal outside China, and a number of other failed projects.

None of this means that go global is not happening, however. Companies like Goldwind have shown they can build successful projects in the US, and are competing to win more. Their traction will grow as they demonstrate a track record across a growing number of international markets.

And Chinese developers are moving ahead with their international plans. In November 2013, Longyuan won the rights to build two wind projects in Northern Cape Province, South Africa, along with local company Mulilo Renewable Energy. The projects, De Aar phases 1 and 2, have a combined capacity of 244MW. In what may become a pattern, United Power has been awarded the turbine contract. Longyuan had previously won rights to develop the 100MW Dufferin wind farm in Ontario, Canada, its first project outside China.

Ultimately, the idea of Chinese turbine manufacturers taking over the world was overblown – GWEC's Steve Sawyer calls it an expression of Sinophobia, and says 'The notion that the wind industry was going to follow a similar pattern to what we've seen in the solar industry, it was never going to happen – that was nonsense, made up by people who don't understand the industry.' But Chinese companies can be expected to have a growing impact on the market in the years to come.

4 Emerging powers: India and Brazil

Another key battleground in the fight to establish wind power as one of the world's dominant energy sources is the big emerging markets. The two most important are India and Brazil, both of which have big populations and steadily growing economies. Both countries need to consistently put on huge amounts of new power capacity to keep the lights on – something which in India sometimes appears to be a losing struggle – and both have turned to wind power in a big way over the last decade.

Following closely are Turkey, Mexico and South Africa, all of which have a significant demand for regular increases in generation capacity, and which have turned to wind in recent years.

Beyond this leading group, there are a whole series of countries which are power-hungry and have begun to see wind as an alternative to fossil fuels. These include Pakistan, the Philippines and Thailand in Asia; Morocco and Egypt in North Africa; the Ukraine; and Chile, Peru and Uruguay in Latin America. There are also a group of countries along a windy East African corridor which are also starting to see wind capacity constructed; these include Ethiopia and Kenya.

There are two features of emerging wind-market growth that are of major significance for the industry as a whole. First, in some of the world's potentially biggest power markets wind is already showing itself capable of competing with and beating fossil fuel and nuclear generation on price. Second, given government moves to ensure local content and local advantage, leading emerging markets are becoming significant wind-turbine producers in their own right.

I want to look at two of the fastest-growing markets; India and Brazil, both of which have markedly different power mixes and policy frameworks for building their industries.

India

On 30 and 31 July 2012, India suffered the largest power cut in history. It affected over 620 million people or around 9 per cent of the world's population, across 22 states in the north and northeast of the country. Around 32GW of power-generation capacity went offline in the outage.

The blackout was the consequence of a power system that is struggling to maintain coherency and keep up with demand. Smaller, but widespread blackouts are a frequent occurrence in India, along with constant 'brownouts' or lack of tension in the electricity supply system. Forced stoppages in factory production are also common.

Just as serious as the perilous state of India's power system is the fact that 300 million people do not have access to power at all, while power supply to many others is intermittent. Losses at transmission, distribution and consumer level are in excess of 30 per cent. A lack of clean and reliable power means an estimated 800 million people use fuel wood, agricultural waste and livestock dung for cooking and other domestic needs, and the resulting indoor air pollution causes between 300,000 and 400,000 deaths per year, as well as other chronic health issues.

With steady – albeit sluggish – GDP growth, population growth and urbanisation, analysts predict that power demand will hit 300GW by 2020–21, requiring 400GW of generation capacity. This means installing something around 200GW of new generation capacity over the next 6–7 years. And even this, of course, would not be enough to ensure that all of the country's population have access to power.

The country's power-generation mix is extremely problematic. India does not have significant amounts of natural gas, leaving it largely dependent on coal. Its own coal reserves are abundant, but of low fuel quality and high ash content. They are also mainly buried under protected tribal and forest areas, and mining projects create fierce local opposition. As a result, India faces a severe coal shortage. Coal imports were a record 138 million metric tons in the 2012/13 fiscal year (Williams and Menon 2013), and India is now the third-largest coal importer in the world after China and Japan. Meanwhile the cost of imported coal is likely to rise to around US$25bn by 2016/17 from US$14bn in 2012/13, frustrating Indian government attempts to reduce its current account deficit.

With its hopelessly inadequate ports and rail infrastructure, the country is literally choking with the stuff. Many power plants typically have reserve coal supplies to last no more than a single day of operations. Coal is damaging both to the quality of life of India's citizens, and to the country's international reputation. By 2012, it had helped to make the country the world's fourth-biggest carbon emitter, and one of the fastest growing, with some analysts predicting that the country's emissions will be on a par with China – if not higher – by 2020. Compared to the average of thermal plants in the EU countries, India's thermal plants emit 50 per cent to 120 per cent more CO_2 per kWh produced.

India started to install its first wind projects in 2000, and the amount of MW installed grew to just over 3GW in 2011, before falling back temporarily to 2.34GW in 2012. The cumulative installed capacity at the end of 2012 was 18.4GW, according to GWEC. The sub-group for wind-power development appointed by India's Ministry of New and Renewable Energy to develop plans for the 12th Plan period, (April 2012 to March 2017) fixed a reference target of 15,000MW in new capacity additions, and an aspirational target of 25,000MW.

Analysts point out that in certain conditions, Indian wind projects have a lower levelised cost of energy than new coal plants. Wind projects are also quicker to develop – they can often be realised in 2 years from initiation to completion, compared to 5–7 years for a new coal plant project. Wind is also scalable and projects can be built with a relatively small capital outlay and then expanded. The combination of these two factors makes wind a powerful tool to solve India's power shortage and stop its carbon emissions from growing.

The main government incentive throughout the early period of India's wind growth has been Accelerated Depreciation (AD), which allowed companies to claim a depreciation benefit of up to 80 per cent of the cost of equipment in the financial year after project commissioning. Although AD was successful in incentivising the first stage of India's development, it created a number of anomalies in the wind market.

With AD, India did not create wind-power developers as in most markets. With a few exceptions, the companies investing in wind projects were so-called 'tax investors'. These were non-energy companies, with small – often 1–3 turbine – projects. The tax investors had no expertise or willingness to get involved in carrying out the range of development and engineering tasks that a wind farm involves. So the other result has been that wind-turbine manufacturers in India have traditionally been developers. This means acquiring the land for the project, which as we will see is a crucial skill with sometimes less than salubrious aspects to it. It also means carrying out a full 'turn-key' contract, providing what EPC services may be necessary, including building access roads, installing the turbines, and building substations and transmission lines if needed.

Analysts have long argued that AD has not helped in creating the scale of projects that is needed to really push down wind-energy prices, while the financial nature of the investment means that investors are not focused on achieving cost-of-energy improvements. It has also meant relatively higher working capital requirements for turbine manufacturers, compared to a supply-only or supply-and-install model that is common in many projects outside India.

Over the most recent period, the Indian wind-power market has been transformed as new frameworks which reward power production rather than capital expenditure have emerged. These included federal generation-based incentive (GBI), as well as state feed-in tariffs and renewable portfolio/green certificate support schemes. As we shall see, the new structures and the phasing out of AD have favoured a new type of large-scale developer whose interest is making profits from wind-power generation. It has also shaken up the market among wind-turbine suppliers.

A new type of market

Indian wind developer Mytrah had many sceptics when it started operations in 2010. Its aim was to set up as an independent, privately financed energy producer that in a short time would become one of India's largest wind-power operators.

Starting as Caparo Energy and carrying out an initial placing of shares on London's AIM stock market, Mytrah has relied on successive rounds of mezzanine finance to carry out its plans, which involve reaching about 1.5GW in operational wind assets in 2015.

It quickly showed the size of its ambition by signing turbine deals for 3GW and 2GW with Suzlon and Gamesa, respectively. It has also managed to steadily build out its portfolio, as well as acquiring other assets. It had around 310MW of wind farms in operation at the end of June 2013, and was expecting to increase this to 548MW by the end of the year. Despite capital spending and paying back loans, the company made a profit of US$3.9m during the first six months of 2013, up from US$2.6m during the same period in 2012.

Chief Executive Ravi Kailas told *Recharge* 'Our numbers seem large, but in relationship to the overall market it is small', citing the total 65GW capacity expected to be installed in India by 2020. 'What we are doing in five to six years' period is do-able. It is a challenge, has not been done before but is not impossible' (Backwell 2011b).

Mytrah is indicative of a new generation of independent power producers set on profiting from Indian wind-power generation. But the expansion of the market has not been smooth.

While AD has been phased out, the initial generation-based incentive (GBI) meant to replace it was widely seen as having been set at an inadequate level and was suspended in April 2012. The government reintroduced it at a higher level in August 2013, but not before the expansion of the Indian wind market had suffered a serious hiatus, with installations in 2012 at 1.7GW, 47 per cent lower than in 2011. Another key area of negotiation are state feed-in tariffs, which are set by the regulatory commissions of each state, and which are not adequate to stimulate wind power in some states. The green-certificate market, which is linked to state renewable portfolio standards, has great promise, but is currently not operational because there are no penalties to hold states responsible for not meeting standards and utilities are not bound to buy clean power.

In 2012, Mytrah and other IPPs like Greenko, Green Infra and Tata Power set up the Wind Independent Power Producers Association (WIPPA) to represent them alongside the Indian Wind Turbine Manufacturers' association (IWTMA) that has traditionally represented the wind-power sector.

WIPPA president and Mytrah Chief Operating Officer Uday Bhaskar Reddy says that the phasing out of AD has produced fundamental changes in the market. 'Now any wind farm that is set up has to be viable on its own, based on quality and sustainability of generation, and these issues may not be of the same interest to manufacturers', he says. 'The IPP sector is the fastest-growing part of the market, and of the 19GW that India has installed, at least 4GW have been built by IPPs' (Backwell 2013a).

WIPPA says that the evolution of the generation-based incentive (GBI) as well as state feed-in tariffs is a key parameter for independent power producers (IPPs). IPPs are also hopeful that reforms will be made to make the green-certificate market function. He says that although the market was only likely to grow by 1.5GW in 2013, rather than the 3GW plus that had been expected, the majority of the power – over 1GW – would be constructed by IPPs.

'Some of the specific challenges facing IPPs are that turbine manufacturers have traditionally carried out turnkey projects for developers, but now IPPs are getting into development and face all the associated challenges, such as getting land', he says. 'Turbine manufacturers have been doing it for a long time, and it will take professionalism and a year or two for IPPs to do it well', Reddy adds.

The rise of Suzlon

Nestled amid a growing cluster of skyscrapers in the traditionally genteel and academically orientated, but now fast-sprawling city of Pune lies the One Earth Centre, the headquarters of India wind-turbine group Suzlon. The 100-per-cent-renewables-powered centre, designed by US-born architect Christopher Charles Benninger, is spread across four hectares and can house 2,300 people.

Indian businessman Tulsi Tanti spotted the potential of wind relatively early and set out to create a global group – but one based in an emerging market – that would bring cheap wind power to the masses. Tanti captured the zeitgeist, with his huge optimism and energy and high-profile engagement with global climate forums and green business groups, becoming one of India's richest men (for a while at least) in the process.

Tanti was managing his family's 20-employee textile company in the early 1990s and said he became interested in energy as he experienced the effects of India's shaky power grid and realised that electricity was his firm's second-highest cost after raw materials. He set up Suzlon in 1995 – entering into a technical collaboration agreement with German company Sudwind GmbH – and adopted a business model where customers would put up 25 per cent of the cost of a wind farm and Suzlon would arrange bank finance for the other 75 per cent. After overcoming initial reluctance, by 2008 over 40 Indian banks were financing wind-power projects for Suzlon clients.

Tanti sold the textile business in 2001 and by the 2005/06 fiscal year, Suzlon was supplying over half of India's wind market. Expansion into the US – Suzlon achieved its first sale there in 2003 – China, Australia, Brazil and other markets was taking place simultaneously. Turbine installations

Figure 4.1 Suzlon's founder and Chairman, Tulsi Tanti (Christopher Bangert/ Danish Wind Industry Association)

passed 1GW in the 2004/05 fiscal year, and by 2012/13 had crossed the 20GW mark.

In 2006, Suzlon paid US$565m to acquire Belgium-based gearbox manufacturer Hansen Transmissions, as it deepened its holdings in the supply chain. In 2007, Tanti engaged in his most ambitious move yet – the purchase of a majority stake in German turbine manufacturer and offshore specialist REpower – which valued the company at US$1.6bn. The battle to win majority control over REpower put Suzlon head to head with French nuclear giant Areva, which was planning its own entry into the offshore wind market, in what newspapers dubbed 'France vs. India for a piece of Germany'.

Tanti won the battle for majority control after five months of intense rival bidding with Areva for REpower shares, and through gaining the support of Portuguese engineering and renewables company Martifer, which had a 23 per cent stake in REpower. The two companies bid 20 per cent over the price offered by Areva in a public offer for shares through a Suzlon-financed special-purpose vehicle, with an agreement that Suzlon would buy Martifer's entire REpower stake two years later. By 2009, Suzlon controlled 90 per cent of REpower, and by October 2011, it had completed a so-called 'squeeze out' process to acquire the share of remaining minority shareholders. The move catapulted Suzlon into the top four global manufacturers, but it was also the major contributor to saddling Suzlon with over US$2bn of net debt.

Tanti's business model was always based on highly optimistic forecasts of global wind-power growth. These were not unrealistic, when the wind industry was clocking up spectacular annual installation figures in the mid/second half of the decade. As we shall discuss in Chapter 6, however, the impact of the global sub-prime crisis on financing, the failure of the Copenhagen climate talks and the overhang in wind-turbine manufacturing capacity, along with a major issue with cracks to rotor blades affecting its turbines between 2007 and 2009, were soon to take a heavy toll on Suzlon's business.

Added to the international factors came a slowdown in the Indian market, as the government moved to phase out AD. Newer incentive schemes such as the GBI or green certificates were either not set at adequate levels or had not been fully implemented. Suzlon's debt problems meant that by 2012 it was facing a major squeeze on its working capital that left it unable to fully execute its extensive project pipeline, and it lost its place as India's number one OEM in terms of annual installations – it was still by far the biggest in cumulative terms – to Wind World (India), the former subsidiary of Enercon in India.

We shall look at the fortunes of Suzlon, and its chances of bouncing back, in Chapter 7.

Not for the faint-hearted – foreign OEMs in India

India's market is not for the faint-hearted. Obtaining land and building permits means negotiating India's state bureaucracy, and this requires a particular local set of skills and experience.

As in many markets, Vestas was a front-runner in India, and was the second-largest supplier in the Indian market, operating through a joint-venture subsidiary with local partner RRB. Fellow Danish company NEG Micon was in first place through NEPC Micon, which was a joint venture with the NEPC group. Like other OEMs in India, both companies engaged heavily in wind-farm development including land procurement.

When Vestas completed its Micon acquisition in 1996, NEPC Micon was split up. Vestas also split with RRB in 2004, and the company consolidated the Micon operations, and moved to a turbine supply-and-installation-only model. It quickly found that it could not maintain market share with this strategy and by the end of 2011 orders had dried up. In September 2012, Vestas announced it was scaling down its India operations, 'to focus on providing value-added service and maintenance to existing Indian wind-power plants' (Ramesh 2012). As we shall see, key staff left in frustration to join rivals.

Now that India's wind market is moving to a new model, with development being carried out by large IPPs, Vestas may have a chance to re-enter the market. But with competition from Suzlon and several other highly aggressive players, it won't be easy.

Enercon India goes 'rogue'

As in China, there are also serious issues with the protection of intellectual property rights as can be seen in the case of German turbine manufacturer Enercon, which found itself fighting a losing battle against a 'rogue' subsidiary.

Enercon made a good start in India, setting up Enercon India (EIL) in 1995 as a majority-owned joint venture. By 2010, it had established itself as the second-largest supplier in India after Suzlon, with its highly regarded 800kW direct-drive turbines.

However, in 2007 a dispute between Enercon and its local partners, the Mehra Group, flared into the open after two years of worsening relations. Even though Enercon held 56 per cent of EIL, the company was in effect controlled by the managing director, Yogesh Mehra, who managed to exclude the parent company's representatives from the board.

A legal case brought by Enercon in 2007, accusing its Indian partners of concealing the company's state of affairs and financial mismanagement, resulted in a court ruling that preserved the Mehra family's control of EIL. From 2007, there was no further co-operation between the two companies. Enercon no longer vouches for the 800kW turbines which EIL continues to build and sell. The Indian company continues to use the parent company's logo and literature on its website, and continues to do well, signing major deals, particularly with Hong Kong-based developer CLP Group for projects like the 113MW Andhra Lake wind farm.

In early 2011, EIL managed to convince a patent court in Chennai to annul 12 of Enercon's patents, arguing that the patents did not represent any genuine innovation. 'In view of this analysis and the finding herein, it is very clear there is no inventive step in the invention claimed in the impugned patent and the invention claimed is obvious to any skilled person in the art', reads one of the rulings seen by *Recharge* (quoted in Backwell 2011a).

Enercon argued that the rulings had major implications for other turbine manufacturers in India. Enercon lawyer Stefan Knottnerus-Meyer says the court based its decision in part on the premise that India's national interest in

Figure 4.2 Gamesa India's CEO, Ramesh Kymal (Ben Backwell, *Recharge*)

developing renewable energies has priority over a company's patent rights. 'In keeping with this argument, almost every new patent could be nullified in the future, in the name of India's development interests', he adds (ibid.).

The rulings left EIL free to carry on producing its turbines and prevented the German group from making any exclusive deal to license the turbine models to another Indian group. On the other hand, it remains to be seen if EIL is capable of designing and building new and bigger turbine models, which are likely to represent an increasing part of demand in India, as the market moves to a new wind-power development model. For now, however, EIL, which is now called Wind World (India), is doing very nicely; in 2012 it replaced Suzlon as the biggest turbine supplier in terms of annual installations.

Gamesa moves in

As we have seen, Vestas' fortunes in India declined when the company moved to a turbine supply-and-installation-only model. Vestas' loss to a great extent turned into a gain for Spanish turbine manufacturer Gamesa, when India boss Ramesh Kymal became the new Gamesa India's CEO, taking key members of the team that had built up NEPC Micon/Vestas with him.

After entering the market in 2010, Gamesa became the third-largest supplier in India in 2011 with a market share of 10 per cent. It built a series of new factories including two nacelle assembly plants in Chennai and blade and tower plants in Baroda, in Gujarat state.

Kymal says that he had been opposed to Vestas' decision to pull out of project development, as well as what he sees as a failure to understand local realities. He says that when he started with Gamesa he was able to persuade the company's global management to start in India with the extremely grid-robust Gamesa G58 850kV turbines to build its brand, rather than a newer, but more sensitive model.

Development activities, which Gamesa carries out worldwide, and the ability of Kymal's team to gain access to land have been key to the company's rapid expansion in the country.

Gamesa is also benefiting from the shift in the market from traditional 'tax investors' to IPP clients and has chalked up some key successes for its G58 and 2MW G97 turbines. Kymal told *Recharge* in Chennai in 2011 that the company went from 100 per cent traditional clients in 2010, to supplying 30 per cent to IPPs this year, and this was expected to increase to 60 per cent in 2012.

In May 2011, Gamesa pulled off a major coup by signing a US$2bn, 2GW supply deal with developer Caparo Energy India (now Mytrah Energy). Under the terms of the deal, Gamesa will provide Caparo with G58 850kW and G97 2MW turbines between 2012 and 2016.

As well as Gamesa, there are also several other companies which are aggressively growing in the market, as well as more well-known players such as GE. One is Indian company ReGen Powertech, which has a licensing agreement with Goldwind-controlled turbine designer Vensys for a highly regarded direct-drive turbine. The company grew fast in 2010–11 and was in third place by new installations in 2012, behind Suzlon and Wind World (India). Managing director Madhusudan Khemka says nearly all ReGen Powertech's sales come from IPPs, rather than smaller 'tax investors'.

Where next?

As we have seen, the phasing out of AD and the suspension of the GBI in April 2012, along with other factors such as worsened financing conditions, meant that the promise of a speeding-up of India's wind development was postponed in 2012 and 2013. But it will come, as India's fossil-fuel-based energy system becomes increasingly problematic.

Gamesa's Ramesh Kymal, who also chairs the Confederation of Indian Industry's renewable-energy council, points out that wind could already compete with conventional power plants if fossil-fuel subsidies were removed. 'This whole change is coming', he says. 'That is why IPPs are becoming much more viable, because the FIT will go up, REC prices are already going up and we are lobbying hard to have the GBI extended and increased. I am looking at a huge market of at least 5–8GW per year of steady growth in India' (Backwell 2011c).

Kymal says the government is looking 'very favourably' at the CII's proposal for India to meet 12 per cent of its power production from wind by the end of the 12th five-year plan in 2017. Adopting the target would imply a huge acceleration in installations.

There are huge challenges, such as implementing central government policies through the states and through local bureaucracy. The physical challenges of constructing wind in India – the lack of decent roads along which to transport wind-turbine blades and towers, and the chronic lack of transmission lines in many areas, which can have crippling effects on both planned and operational projects – should not be underestimated, but they can be overcome with investment and political will.

Top 10 wind-turbine suppliers 2012/13

Wind World (India) formerly Enercon (India) 454MW
Suzlon 415MW
ReGen Powertech 273MW
Inox 264MW
GE 122MW
Gamesa 100MW
Vestas 34MW
RRB 14MW
Kenersys 10MW
LSML 9MW
Source: IWTMA

Brazil

Brazil takes off

Brazil shares India's high growth potential, with steady expansion in economic outcome and power demand forecasted, boosted in the short term by the expectation of a 'Brazilian decade' with highlights including the country's hosting of the 2014 football World Cup and the 2016 Olympics.

Like India, Brazil faces a struggle to keep the lights switched on, although its power matrix could not be more different. Around 80 per cent of Brazil's power comes from hydroelectric power, while thermoelectric power from gas, fuel oil and coal makes up just over 16 per cent. Nuclear power makes up 2 per cent of the total supply.

Brazil's government sees developing clean energy as essential to continuing to project itself as a global leader in terms of 'soft power' as it pushes for a place at the top table of world politics (see Backwell 2013b). Traditionally the availability of hydrocarbons has been limited, and this allowed Brazil to become a global pioneer in biofuels in the 1970s. The arrival of a boom in 'pre-salt' ultra-deep-water oil exploration has not brought any immediate large increase in flows of natural gas to the mainland, and it is still debatable how big such a flow will be in the future. Hydroelectric power will remain the mainstay of Brazil's power system. While this is positive in terms of carbon emissions, it is difficult to add large amounts of new power, given that hydro projects require huge capital investments, long-term planning and have major impacts on the environment and local populations – big projects such as Belo Monte have caused huge public controversy. The hydropower supply can be sharply curtailed if there are several consecutive drought years, as there were in the run up to the 2001–02 energy crisis, which Brazil overcame without blackouts only by reducing consumption by 20 per cent over an eight-month period. Hydro-generation relies on highly seasonal patterns of rainfall, and this can bring Brazil's power system close to the brink.

Government policymakers who were once sceptical about wind power have discovered that wind power is highly complementary to hydropower, with Brazil's winds blowing most strongly during the winter dry period. A report by GWEC in 2012 says:

> Hydro and wind power are perfect partners in Brazil. Not only are the country's windiest areas located conveniently close to demand centres, but in addition, the variable nature of wind power is best accommodated in a highly flexible system such as one dominated by hydropower. Furthermore, wind power can help alleviate some serious energy security concerns in Brazil, especially during the dry winters.
>
> (GWEC 2012)

One of the consequences of the 2001–02 energy crisis was the creation by the new PT administration of an auctioning system in 2004, as part of a series of reforms aimed at ensuring that adequate amounts of new power would be brought online. As part of the reforms, the government also created a new company, EPE, tasked with long-term energy planning and deciding – among other things – how the auctions should be designed and which energy sources should be included, based on the criteria of energy supply security and cost-effectiveness.

Brazil has extremely good wind resources. In large parts of the northeast of the country, the wind is constant – industry officials compared it to a vast warm hair dryer – and this has allowed Brazilian wind projects to clock up some of the highest average capacity factors in the world, at in excess of 50 per cent. In addition, wind capacity can be quickly brought online, as industry officials explained during a series of meetings in the early part of the decade, as they struggled to get wind taken seriously in Brazil.

The government established the legislation for the first support system for wind called PROINFA during the administration of Fernando Henrique Cardoso, although this was not implemented until the first Lula government by the then Energy Minister – now President – Dilma Rousseff. Government officials continued to be fairly sceptical, however, and the breakthrough moment did not come until spring of 2009, when wind-industry bodies organised a trip to the then booming wind market of Spain for key Brazilian politicians and government officials, including members of the renewable-energy committees in both the upper and lower houses of the Brazilian Congress. 'Seeing the scale of the industry, and most importantly the Red Electrica control room, from which the whole Spanish power system is controlled, caused cascading epiphanies in the minds of Brazilian officials and politicians', says GWEC's Steve Sawyer (personal communication).

Among those on the trip was EPE President Maurício Tolmasquim who organised the first capacity auction later in 2009, and who says that 'wind's moment' has arrived. 'I was never against wind power', Tolmasquim told *Recharge* correspondent Milton Leal in his offices of the 11th floor of an imposing commercial building overlooking Rio de Janeiro's beautiful Guanabara Bay.

> Inside the Energy and Mines Ministry, I was one of the most favourable. During the time of [the alternative-energy incentive scheme] PROINFA, I defended the need to give a chance for wind. But at the time price was a barrier ... The planner can't have a passion for, or hate a form of energy. He has to have a rational view. Above a certain price level, I couldn't justify contracting wind.
>
> (Leal 2013a)

Today this is possible, as Brazilian wind projects have been bidding for capacity at prices that are arguably the world's lowest and beating fossil-fuel projects on equal terms along the way.

At the December 2012 A-5 auction ('5' being the time period in years within which the new power needs to come online) Brazil awarded contracts to four developers to sell 281.9MW of capacity from ten wind farms due to be completed in January 2017. The average rate was R$87.94/MWh (US$42.2/MWh), with the lowest prices being R$87.77/MWh. Average prices were 12 per cent cheaper than the R$99.58/MWh average price in August 2011, which at the time was considered by some to be the lowest price for wind power in the world. These projects compare with the cheapest power purchase agreements (PPAs) in the US, which come in the range of the mid-US$50s per MWh if the government's production tax credit is factored in.

Brazil's highly competitive auction system has created a number of dynamic local developers, whose business models are extremely aggressive. The most impressive is Renova.

Run by 41-year-old former McKinsey consultant, Mathias Becker, Renova has expanded at breakneck pace. When Becker was appointed at the start of 2012, Renova had 110 employees and zero wind installations. By September 2013, it had 294.4MW installed (although not grid-connected), 1.29GW of PPAs, a 12GW pipeline across Brazil, a 280-strong workforce and an average annual growth rate of 30 per cent mapped out until 2017 (based on PPAs signed before August 2013).

Becker became involved in Renova when one of his consulting clients, state-owned utility Cemig, was looking for a way to enter the renewables sector. Becker suggested investing in Renova – which was founded by Ricardo Delneri and Renato Amaral, and which at that time had won 462MW in national tenders. Cemig bought a 25.8 per cent stake in the company for R$360m (US$158m) through its subsidiary Light. In August 2013, Cemig announced that it was pouring R$1.41bn (US$ 0.64bn) into the developer via another subsidiary focused on generation and transmission, joining Light and a third stockholder, RR Participações, in a block controlling at least 51 per cent of Renova's shares. Six months into Becker's reign, the Brazilian national development bank, BNDES, bought a 12.2 per cent stake in the company for R$260.7m (US$117m).

Having both BNDES and Cemig on board has paid dividends for Renova, significantly reducing the cost of its financing. The stock market has also shown more confidence in the company – Renova's share price has risen 200 per cent since the initial public offering in 2010.

The force driving Renova's growth is the extraordinary winds of northeastern Brazil. At Renova's Alto Sertão complex in Bahia, Brazil, the winds blow in from the east at an average eight metres per second, with a daily lull from around 11am to 4pm that makes it safe enough to fly a helicopter – or install a turbine.

Capacity factors are a spectacular 45–50 per cent and there is a proven resource of 6GW across the 11,000sq km that Renova has leased from local landowners. An excellent internal rate of return on the 294.4MW first phase is set to rise for the second and third phases. It is little wonder that the company refers to Alto Sertão as its 'gold mine'. 'We want to reproduce the same Gold Mine effect [in other parts of Brazil], having Gold Mine 1, 2, 3, 4, etc.', Becker told *Recharge* (Leal 2013b).

But not everything has gone Renova's way. Through no fault of its own, it was still not producing any electricity at the time of writing. The substation and transmission line that will flow Renova's wind energy to the grid have yet to be built by state-owned utility Chesf at the time of writing and are more than a year behind schedule, and it seems likely that the 386MW second phase of the project will be ready – in March 2014 – before the transmission lines. Renova is being paid by the state for the electricity it would have produced, but this is hardly of benefit to Brazil's consumers and the company has to pay for oil-fired generators to spin the turbines' rotors for maintenance purposes.

Sergio Marques's big bet

An aggressive development can sometimes spill over into what some analysts consider as 'reckless', as in the case of rival developer Bioenergy, which is the creation of controversial Brazilian businessman, Sérgio Marques.

Marques – like Becker relatively young in his late 30s – has almost single-handedly grown his company from a one-man band to a multi-million-dollar business using little more than his own creativity, drive and powers of persuasion. His projects have been the lowest of the lowest priced in Brazilian wind auctions, with 201MW of Bioenergy projects bid at the record R$87.77/MWh (US$40/MWh) price in the December 2012 A-5 tender.

In 2002, Marques was an executive at a subsidiary of ABB, which was then trying to break into the Brazilian wind sector. When the Swiss–Swedish technology giant pulled the plug on the unit, Marques set up Bioenergy and took all the projects he had been developing for ABB with him.

Between 2005 and 2009, Bioenergy signed about 400MW of long-term PPAs, either through national tenders or on the open market. The money he gained from selling these PPAs was used to fund the company's only two operational wind farms, the 14.4MW Aratuá 1 and the 14.4MW Miassaba 2 – the latter being the first in the country to deliver power to the unregulated market. By late 2013, Bioenergy had signed around 800MW of PPAs.

Marques is able to make extremely aggressive bids for capacity because the success of his projects are predicated on them being built ahead of time. For the R$2.5bn (US$1.1bn) Paulino Neves wind project, for example, Marques bet that the R$1.4bn (US$0.6bn), 351.9MW first phase of the 640.9MW complex could be built ahead of time, so that the turbines can reap huge amounts of money on the unregulated market before part of the PPAs signed in the regulated market start in March 2014. Marques was so confident that he has already sold some of this pre-PPA output and if the annual national spot price stays at about R$120/MWh (US$54/MWh) in 2014 – a figure considered conservative by consultants – Bioenergy could see a R$300m (US$135m) profit in 2014 and millions more in 2015.

Marques thinks the average spot price will be closer to R$170 (US$77), which would make him huge profits. However, if the project falls behind schedule, Marques faces ruin. The penalties for those who sell but don't deliver energy can be astronomical – Bioenergy would have to buy in the power it was supposed to supply at spot prices, which can fluctuate wildly from R$15 (US$7) to R$780 (US$350). At the time of writing, it is by no means certain that the wind complex in the northeast state of Maranhão will be ready on time. The state currently has zero wind capacity and Bioenergy is building the required transmission lines itself.

There is considerable scepticism around Marques's business model. A top executive of a big European developer active in Brazil says that Marques will eventually have to try to sell his portfolio, although at the prices he has bid it is not clear who would be the buyer. Immediately after the December 2012 A-5 auction, Elbia Melo, head of the Brazilian wind-energy association, ABEEólica, claimed that the price Bioenergy agreed was unrealistic and did not represent the real costs of wind generation. Marques countered that Melo was not taking into account the near 5 per cent annual inflation attached to the PPA or his intention to sell part of the output on the free market.

Melo subsequently said that Marques can be compared to Eike Batista, the Brazilian business magnate who was the world's eighth-richest man in 2012 after amassing a US$34bn fortune from mining, oil and other ventures. She may have chosen her words carefully – Batista is now effectively bankrupt, after falling oil production prompted an investor sell-off in his businesses. Time will tell if Marques, with his innate ability to make money and his fondness for risk-taking, suffers a similar fate.

The OEMs pile in

The creation of a dynamic Brazilian market, combined with demanding local content conditions, has been extremely successful in drawing in a large number of international turbine suppliers.

The local content rules operate through the national development bank, BNDES, which is practically the sole local source of financing for wind-power developers and offers extremely attractive – sometimes negative if inflation is taken into account – interest rates. In order to qualify for financing for bids in Brazil's power tenders, developers' designated turbine suppliers need to meet the BNDES conditions. These conditions in turn have become more complex, moving from an overall percentage of all the turbine equipment used in a project, to a 'through the turbine' approach that requires manufacturers to meet different percentages in several different areas of the turbine.

The first movers in Brazil were Germany's Enercon through its Wobben Windpower subsidiary, Argentina's IMPSA, Suzlon, GE, Alstom and Vestas.

Wobben's turbines are highly admired in the Brazilian market, while IMPSA has pursued an aggressive growth model based on turbine sales and participation of its own development arm, Energimp, in the power tenders. IMPSA had all its manufacturing based in Brazil from the start, giving it a significant advantage in meeting the second phase of local content rules. The company has an impressive 1.26GW of turbines installed and on order in Brazil; the country's largest wind-turbine production capacity (1.4GW) and formidable political contacts within federal and regional governments

IMPSA Vice-President José Luis Menghini says it has poured US$250m into its Brazilian infrastructure so far, with a further US$60m being spent on a new 400MW nacelle factory in the southernmost state of Rio Grande do Sul, which would also supply the Argentine and Uruguayan markets. IMPSA's existing 1GW nacelle plant at the Port of Suape, in the northeast state of Pernambuco, is operating at full throttle, as the company has to deliver most of a 900MW-plus backlog by the end of next year.

Since 2008, IMPSA has won 1.26GW of orders in Brazil – a 15 per cent slice of the market – but most of that (803MW) was ordered by Energimp. The remaining 460.5MW was split between state-owned generators Chesf (180MW) and Eletrosul (234MW); Tecneira, a developer of mainly small-scale projects, owned by Spain's ACS group (42MW); and Australia's Pacific Hydro (4.5MW).

IMPSA has long been regarded as a manufacturer of quality equipment with its direct-drive machines licensed from Goldwind-owned Vensys, although it has had some problems. It may also find it very challenging to compete with the big balance sheets and international experience of companies like Alstom and GE in the turbine market on the one hand, while competing in a very low-price development market through Energimp is also far from straightforward.

In recent tenders, the most noticeable contract winners have been French-owned Alstom – a relative minnow up until now in terms of global sales but one with big ambitions – Spain's Gamesa and the US's GE.

Alstom wins big with Renova

The size of the opportunity that Brazil represents for manufacturers became clear in February 2013, when Alstom signed a €1bn (US$ 1.4bn) deal to supply at least 1.2GW to Renova. Alstom described the deal as its 'most important' ever in onshore wind. It has built a second factory in the south of Brazil, to complement its plant in Camaçari, Bahia, where it has introduced a double shift, increasing capacity to 600MW annually.

Alstom's tie-up with Renova caps a run of contracts won since mid-2010, when the manufacturer was tapped by São Paulo-based Desenvix to supply 57 of its 1.67MW ECO86s for the Brotas development in Bahia. In 2011, Alstom signed a €200m (US$275m) agreement to supply and maintain 41 ECO86 machines for a trio of Brasventos wind farms in Rio Grande do Norte state. In 2012, Alstom signed deals with Odebrecht Energia – a 108MW order to kit out four wind farms in Rio Grande do Sul with 40 ECO 122s; Casa dos Ventos – a €230m (US$316m) deal to supply 68 ECO122s in Rio Grande do Norte; and a pair of contracts with infrastructure group Queiroz Galvão for Eco122s for two complexes in the same state. 'Latin America, especially Brazil, will be one of the centres of wind development in the coming years', says Alstom's Brazil country manager Marcos Costa. 'Brazil will also be the centre of development and production for wind farms in other Latin American countries' (Backwell 2013c).

The Renova deal is particularly notable in that the Brazilian developer had been working previously with rival OEM, GE. Officials from Alstom say the key to winning the contract was offering a range of turbine options that could be tailored to the wind conditions of each individual site. Alstom and Renova have set up a joint committee, and are working together to optimise Alstom's turbines for the idiosyncratic winds of the Caetité region. 'We have bet on the best business plan, not on the cheapest machine', says Becker (Leal 2013b).

Because of the Alstom deal, the manufacturer was able to convince its blade supplier, Tecsis, and tower maker Torrebras to build factories in

Bahia – dramatically reducing transportation costs for Renova. Blades for the GE turbines currently take 25 days to travel the 1,500km from São Paulo state. For the third phase, the blades and towers will travel less than 700km. 'Renova is almost a market in itself for us', says Marcos Costa, Alstom's Brazil country manager. 'Brazil is the bigger market and within it, Renova is a super market' (ibid.).

Gamesa bets on Brazil for growth

As we will see in Chapter 7, Spain's Gamesa spent the early part of the current decade struggling to reinvent itself, after orders from the Spanish market dried up, and sales to its shareholder and partner Iberdrola were no longer enough to guarantee growth.

In a move to capture sales from fast-growing emerging markets, Gamesa made an aggressive entry into India's wind market in 2010–11, and was simultaneously making moves in Latin America, an area in which many Spanish companies have traditionally seen themselves as having an advantage.

In Brazil, Gamesa set up a manufacturing plant in Bahia state, and was a big winner in the new power tenders, winning over 600MW of orders between 2009 and 2011, using a strategy of competing with ultra-low margins to gain critical mass in the market place.

By the end of the third quarter of 2013, Latin America was the destination for fully 51 per cent of total sales, with the bulk of this coming from Brazil. Gamesa was one of the first companies to confirm that it had complied with the latest BNDES local content rules and obtained FINAME approval.

In the August 2013 reserve auction, companies that had registered to use Gamesa turbines won at least 462MW of capacity, making it the single biggest winner, and the company announced plans to expand its Bahia plant to increase output.

How many companies does Brazil need?

Brazil's combination of competitive auctions and local content rules through the BNDES can be considered a success in terms of fostering the fast growth of the wind sector at competitive prices and building up a new manufacturing industry. But the policy has perhaps been too successful. At a time when growth opportunities in the global wind-power market were relatively scarce – as we have seen, Western companies have been largely shut out of China, the most dynamic market in the period – OEMs have piled into Brazil.

As of early 2013, there were at least ten companies manufacturing or planning to manufacture in Brazil: Enercon/Wobben, IMPSA, Vestas, Gamesa, Alstom, GE, Acciona, WEG, Siemens and Suzlon. For a market that is likely to be around 2.5GW per year – in a fairly optimistic estimate – this is clearly too many to sustain. Competition and pressure on margins have been intense.

Gamesa's country manager Edgard Corrochano says, 'With a 2GW-a-year market, the most logical thing would be for three to four – most likely three – manufacturers to stay in the market, and we will be one of them.' To make the grade, he believes 'You need decent volumes, and we have that, along with maybe two more' (Backwell 2013d). He predicts that by 2015, consolidation in the market will have played out.

The new FINAME regulations will force companies to make significant new investments in manufacturing, and companies are making hard calculations whether to simply fulfil existing orders or stay the course and invest to be able to take part in future tenders. 'Meeting [the requirements] is hard work, but we have been working on it continually for three years', Corrochano says (ibid.). For potential competitors, 'It's not easy to come in and do now what I have already done', he says, adding that those companies that want to stay in Brazil 'are going to have to make big investments' (ibid.).

As of late 2013, Wobben, IMPSA, Alstom, GE and Gamesa said they had complied with the new regulations, while Acciona said it was on course to do so. Vestas, one of the leaders in terms of constructing wind farms following success in the early tenders, announced in June 2014 that it would invest R$100m ($43.5m) to ramp up manufacturing capability in Brazil and meet the FINAME requirements.

Can Brazil's wind power stay cheap?

FINAME 2 is likely to have the effect of driving consolidation in the Brazilian turbine market. But it is also driving up turbine prices and threatens to undermine the price advantage that the Brazilian wind sector has enjoyed until now.

In 2013, turbine prices increased by around 20–25 per cent, because of the stricter local-content rules, helped by a good level of power demand at the latest tenders. Several turbine companies are having to reassess their involvement in the market given the level of competition and renewed pressure on margins, while there has also been a lot of grumbling over the onerous level of inspections and declarations required by the new rules, as well as the difficulties of obtaining some components locally at realistic prices. On the other hand, some of the players doing well in Brazil point to the fact that by producing more components locally, the industry is less exposed to exchange-rate fluctuations.

Alstom's Senior Vice-President for wind Alfonso Faubel told me in Rio de Janeiro that his company fully supports the new BNDES regulations, and that he is confident that the cost of producing turbines will come down in the medium term.

'The key thing is scale and an automotive-industry approach to production', said Faubel, who added that in terms of standard hours, the French-owned turbine manufacturer expects its Brazil nacelle assembly to reach very competitive levels in global terms in 2014.

Figure 4.3 Alstom's Senior Vice-President for Wind, Alfonso Faubel (EWEA)

Brazilian President Marcos Costa says 'the policy of [national development bank] BNDES is to localise production. In the short term, some components can become more expensive, but this is a learning curve, and we fully support the policy' (Backwell 2013c).

There are still many question marks over how big a chunk of Brazil's energy demand wind is capable of, or will be allowed to take. The country seems set for a big expansion in its oil and gas production through the exploitation of Brazil's so-called 'pre-salt' hydrocarbons deposits in the Santos Basin. State oil company Petrobras is still one of Brazil's most powerful lobbies, and depending on export markets and international gas prices, increasing flows of gas – yet to materialise, it is fair to point out – may result in increasing pressure to build new thermal plants, supported by the traditional arguments around the need for 'base-load power'.

Or not. Every year that Brazil's wind industry demonstrates to policymakers and the public that it is capable of rapidly adding power capacity, at competitive prices and without the huge capital expenditure costs and environmental risks that the pre-salt involves, will make it harder for wind to be pushed back to the margins of Brazil's power system.

5   The offshore frontier

By placing turbines in the sea, we can reap significantly higher levels of energy because of the strength and consistency of winds offshore. The resource is such that a relatively small area in the Northern European seas area could – theoretically – create enough energy to supply the entire world with electricity. The countries around the North Sea – Denmark, the UK, Germany and others – have been the first to take advantage of offshore wind's possibilities, but there are good resources offshore from the US to China, Japan, Korea and India.

Where offshore wind is developed is a function of economics, and countries need both good resources and a compelling reason to build wind power in the sea. Some, like the US, have good offshore wind resources, but there are abundant development opportunities for cheaper onshore wind. For others, such as the UK, offshore wind has become one of the only ways apparent to add zero-carbon power in large quantities – and relatively quickly – to replace its ageing generation base. The UK, which has become the undisputed leader in the sector, is fortunate to have probably one of the greatest resources of any single country in the world, with industry officials touting it as 'the Saudi Arabia of offshore wind'.

Harvesting this wind resource constitutes possibly one of the greatest engineering challenges that mankind has faced. Installing turbines in deep, rough waters means that they need to sit on stable foundations, and it is these that constitute one of the biggest costs and technical challenges. The other challenges are: installing the foundations and turbines in constantly changing weather conditions; ensuring that turbines actually work, and work reliably over long periods of time, withstanding constant strong winds, storms and corrosion; and connecting often offshore wind farms to onshore electricity grids, requiring the building of offshore substations and the laying of large array cables on the seabed.

These challenges have inverted the economics of wind-farm development. In onshore projects, the wind turbine typically makes up 65–70 per cent of the cost of a project, while in a typical offshore project the turbine may constitute only 40–45 per cent of the cost, due to the high cost of the foundations and installation. The installation process and the cost of sub-structures are relatively inelastic at their lower range and so it makes sense to try to increase the amount of energy produced by each turbine. The result is a steady increase in turbine size, taking us to the 8MW Vestas turbine, which we mentioned at the beginning of the book, and beyond, with several companies now working on 10MW designs.

Beginnings

As we have seen, Henrik Stiesdal supervised the construction of the world's first offshore wind farm in 1990–91, 2.5km off the coast of Denmark at Vindeby. The wind farm was developed by Danish utility association, Elkraft, that later merged into DONG Energy, a company that would become one of the leading forces in the sector.

Until 2000, growth in offshore wind was slow, with development taking place on a small number of near-shore projects in Danish and Dutch waters, with turbines of less than 1MW capacity. The Middelgrunden project in Danish waters was the first large-scale project with twenty 2MW turbines, while seven 1.5MW turbines were connected to the grid off Utgrunden in Sweden the same year.

Since 2001, offshore wind has been steadily picking up momentum and the share of offshore in total annual wind-capacity installations has been growing, making it an attractive growth area for larger turbine manufac-turers operating in a crowded onshore market. In 2001, the 50.5MW of capacity installed represented 1 per cent of total new European capacity. In 2012, offshore wind installed 1,166MW, representing 10 per cent of the European wind market total of 11.895GW. By June 2013, over 6GW of off-shore wind had been installed.

The Vindeby project featured eleven 450kW turbines, giving it a total capacity of 4.95MW. Just over two decades later, in 2013, partners DONG, E.ON and Masdar were inaugurating the first phase of the London Array project, with 630MW of capacity, made up of 175 Siemens turbines each rated at 3.6MW. By 2012, the average turbine size had increased to 4MW, and this is expected to increase sharply over the next few years as a number of 'next generation' turbines come to market.

While Denmark got things started in offshore wind, as it did in the mod-ern wind market as a whole, and scaled up the size of its wind farms, the amount of potential it offers to developers and turbine manufacturers is dwarfed by its North Sea neighbour, the UK.

The prize: UK offshore wind

The UK is now the undisputed leader in the offshore sector, with 3,461GW of capacity, versus Denmark's 1,274GW, Belgium's 453MW and Germany's 385MW.

The UK is estimated to have over a third of Europe's offshore wind resource, and estimates give it a total theoretical potential in all waters of 120GW using only areas with water depths of less than 50m. The first offshore wind farm, Blyth Offshore, was built under the Non-Fossil Fuel Obligation (NFFO) and commissioned in 2000. It consisted of two 2MW Vestas turbines and was developed by a consortium including E.ON and Shell Renewables.

In 1998, the British Wind Energy Association (now RenewableUK) held discussions with the government and the Crown Estate, which owns almost all the UK coastline and its seabed. A set of guidelines were published that were intended to allow companies to develop wind farms of up to 10km² and 30 turbines, to allow developers to gain experience. In what became known as UK Round 1, 17 applications were given permission to proceed in April 2001.

The first Round 1 project was North Hoyle, completed in 2003, and since then ten more have been completed, with a total of 1.1GW. In December 2003, the Crown Estates announced the results of Round 2, with fifteen projects awarded and a total capacity of 7.2GW, with the biggest being the 1.2GW Triton Knoll area. Two Round 2 projects – Gunfleet Sands and Thanet – were completed in 2010, with several others following over the next couple of years, including London Array in early 2013. In May 2010, the Crown Estate gave approval for extensions to seven Round 1 and Round 2 sites, to allow the creation of another 2GW of capacity.

In the meantime, the Crown Estate had launched a third round of allocations in June 2008 that would dwarf anything that had come before, with potential for up to 33GW in nine zones. Bidding closed in March 2009 with over 40 applications, and with multiple applications for each zone on offer. The successful bidders were announced on 8 January 2010. The first planning permission applications for Round 3 zones were submitted in 2013, with a number of big power companies including Iberdrola, EDPR, SSE, Statoil, Statkraft RWE, E.ON, Mainstream/Siemens, Eneco and Centrica winning areas in different consortia, along with one small but highly aggressive specialist developer, SeaEnergy Renewables.

In parallel, the Scottish and the Crown Estate also called for bids for potential projects in Scottish territorial waters. Although these generally have deeper waters than in the other UK sites, seventeen companies submitted bids, and the Crown Estate signed exclusivity agreements for 6GW of sites with nine companies. Subsequently, five projects have been granted agreements for lease and several of these are now close to gaining planning consent.

The amount of potential GW on offer, and the Crown Estate's highly proactive role in trying to take forward development of the different zones, had large utilities and turbine manufacturers salivating. During the last two years of Gordon Brown's Labour Party government (2008–10), a number of turbine manufacturers including Vestas, Siemens, GE, Gamesa and Clipper had identified sites for manufacturing for new turbines meant to address the coming demand for UK projects.

Figure 5.1 The giant London Array offshore wind farm in the Thames Estuary (Mark Turner/London Array)

Clipper began building a 4,000m² facility on Tyneside to manufacture the blades for its giant 10MW Britannia project in 2010. The same year, GE announced plans for a UK manufacturing facility with a planned £100m (US$169m) investment to create as many as 2,000 jobs. Siemens had announced plans for a £210m (US$354m) facility to build its new 6MW direct-drive turbine in Hull, while Vestas was planning a 70-hectare site in Sheerness in southeast England. Spain's Gamesa and France's Areva subsequently announced plans for manufacturing around the port of Leith in Scotland, while several other companies, including Alstom and REpower, were engaged in scoping sites for their facilities. A number of non-European companies, including Mitsubishi, Samsung, Hyundai, Doosan and XEMC, also developed plans to deploy in the UK. Most of the talk was around a lack of available sites for testing new turbines and building manufacturing facilities, as well as actual and potential supply-chain bottlenecks, particularly in the areas of offshore cables and cable laying and in installation vessels.

The onset of harder times for utilities, followed by the new Conservative government's plans for wholesale reform of the electricity market (known as EMR), which involved scrapping the UK's highly successful Renewables Obligation and replacing it with a system based on contracts for difference (CfD), led to a significant slipping in the previously foreseen timelines for project development of Round 2/Round 1–2 extensions/Round 3 and Scottish Territorial areas. Greater uncertainty coincided with the onset of financial hard times for several turbine makers in particular, and most of the early plans for the creation of a UK offshore turbine industry were cancelled or delayed.

Clipper cancelled its plans for Britannia in August 2011 after UTC pulled the plug on the project. Although it never formally cancelled its plans in the UK, GE pulled back from any large-scale deployment of its planned 4.1MW direct-drive turbine after having second thoughts about both its platform and the market. Korea's Doosan fell by the wayside in early 2012.

Vestas was in crisis by the end of 2011, and, although it received planning permission for its Sheerness site in May 2012, it cancelled its plans a month later in June, leading to a messy dispute with the site owner, Peel Ports. The decision was linked to the squeeze on capital expenditure at the Danish company and the postponement of the deployment of what was then its planned 7MW turbine, as well as to its changing expectations over UK market growth.

Siemens, which had taken the lion's share of the Round 2 contracts and was quietly winning the contracts for most of the earliest-stage Round 3 projects, maintained its plans for its Green Port Hull facility, after having gained planning permission in May 2012. In March 2014, Siemens finally gave the go-ahead to start work on the factory after long negotiations with the UK government. Renewable UK chief executive Maria McCaffery said:

> It's the green-collar jobs game-changer that we've been waiting for. Attracting a major international company like Siemens to the UK, creating 1,000 jobs manufacturing turbines at two sites in Yorkshire, proves that we can bring the industrial benefits of offshore wind to Britain.
>
> (Lee 2014a)

Meanwhile, Spanish turbine manufacturer Gamesa had also been forced to slow down its offshore programme in 2012, relocating its 5MW prototype project from Virginia to the Canary Islands and putting its plans for a 7MW turbine to be engineered in the UK on the back burner. It did maintain interest in a site around Leith, however. Others, like Areva and Alstom, also slowed down their plans. The French turbine makers found that they had significantly more flexibility than some of their competitors, having won big in the French offshore tender in April 2012. The tender obliged both companies to create industrial facilities in France, and they will soon be in a position to supply some UK demand, particularly in the southern part of the country, from their French factories.

As of the end of 2013, UK offshore ambition in the next few years has been significantly scaled back, with the most optimistic forecasts giving a total of 13–16GW of capacity by 2020, compared to previous – rather over-ambitious even in the best-case scenario – estimates of up to 32GW. The final details of the UK's new support scheme were due to become clear at the end of 2013.

The old debate around test and demonstration sites is virtually irrelevant, as a number of manufacturers have either pulled out or decided to engin-eer and test their new turbines outside the UK and the consensus among analysts at the time of writing is that the best the UK can hope for in the immediate future is two manufacturing plants: Siemens' plant in Hull plus one more in Scotland or northeastern England. North of the border, manu-facturers Areva, Gamesa and Samsung are still keen but watching closely the amount of capacity that will emerge and the timing of projects – as well as what Siemens will snap up. In the medium term, however, the game is very much on. The UK needs to lower its carbon emissions, keep the lights on and replace its past-their-sell-by-date plants, and large-scale alternatives to offshore wind are limited. Meanwhile the wind industry is making signifi-cant steps towards making cost reductions, which should eventually make offshore wind competitive with conventional power sources.

Japan – the next big offshore market?

Outside Europe, one of the most promising markets for offshore wind is Japan, whose energy system is being transformed in the wake of the Fukushima Daiichi nuclear accident in March 2011. Analysts estimate that as much as 90GW of both fixed and floating turbines could be installed within the next ten years, and Japan's floating wind demonstration projects have given it a world-leading position in this emerging segment.

When Japan scrambled to shut down its 50 main nuclear power stations after the meltdown at Fukushima Daiichi in March 2011, it was left with a gaping 30 per cent shortfall in the production capacity needed to meet national electricity demand.

In the period between the accident and the end of 2013, the country has spent around 15tr yen (US$145bn) on imports of LNG (liquefied natural gas) to make up the output deficit, a running cost understatedly deemed 'unsustainable' by Shinzō Abe's government.

At the time of writing, there is still considerable uncertainty about how much of Japan's nuclear capacity will be brought back on line. The govern-ment's National Policy Unit recommendations have suggested a range for nuclear power's contribution to the energy mix of 20–25, 15 or 0 per cent for the medium term, while public opinion remains divided.

Offshore wind is seen as the likeliest utility-scale energy resource to sub-stantially reduce the country's reliance on nuclear. Japan has a rich wind resource off both its Pacific and continental coasts, estimated by the Ministry of Environment (MoE) to be equal to a theoretical capacity of 1,600GW.

The island nation's narrow continental shelf means as much as 80 per cent of this potential lies over deep water, in depths greater than 50m, which is beyond the reach of conventional fixed-foundation technologies such as monopiles, jackets and concrete gravity-base solutions (CGBS).

The Japanese Wind Power Association estimates an upper-range figure of 519GW in floating offshore wind capacity, but a more realistic 141GW is thought to be economically harnessable in the medium term. To achieve this, the industry will need to be backed by a new feed-in tariff (FIT) to ensure that investments are economical. This seems set to materialise sometime in the first half of 2014.

Four demonstration projects have so far been launched by the Japanese government through its R&D arm, the New Energy and Industrial Technology Development Organisation (Nedo), to test and fine-tune the leading technologies. At the heart of the initiative is Fukushima Forward. Backed by more than US$230m from the MoE, the 11-member consortium led by industrial giant Marubeni, and including Mitsubishi Heavy Industries, Japan Marine United and Shimizu, is studying the viability of a range of floating wind-power concepts through a cluster of four experimental units moored within sight of the stricken Daiichi nuclear power station.

During the summer of 2013, a 2MW Hitachi downwind wind turbine on a Mitsui semi-submersible foundation – dubbed Mirai – was floated out for grid connection, along with a 25MW 66kV floating substation, Kizuna, and installed in depths of 120m in the Pacific Ocean. This first phase is now up and running. Phase two will begin in summer 2014, with the installation of two 7MW Mitsubishi SeaAngels – the first on a V-shaped semi-submersible and the second on an advanced spar foundation. The pilot is the only multiunit floating development off Japan and may ultimately lead to around 1GW of installations in the waters off Fukushima by 2018.

Off the westernmost tip of Japan, near Nagasaki, a second pilot project was brought online in 2013 by a consortium including Fuji Heavy, Toda Construction, Fuyo Ocean Development & Engineering and Kyoto University. The Goto demonstrator – a Hitachi 2MW on a novel 'super-hybrid' concrete spar hull moored in some 100m of water – replaced a scaled 100kW version of the turbine that had been floating on location since June 2012.

The extent to which offshore wind (along with other technologies like solar PV) fully or partially replaces Japan's nuclear capacity depends on the government's Basic Energy Plan, which is still in draft and is yet to be approved. Theoretically, replacing all the current nuclear capacity would require 90GW of offshore wind, with a further 60GW to take the place of the previously planned nuclear expansion. In other words, 150GW of installations could be the upper 'blue sky' range of offshore installation expectations over the next 16 years.

However, at least some of Japan's nuclear power stations are likely to be restarted – how many remains a moot question. It is also worth remembering that Japan is targeting more than 28GW of solar PV installations by

2020 and 53GW by 2030. Nevertheless, the offshore opportunity is likely to be considerable.

The Fukushima Forward and Goto pilots have been propelled into being in a very short span of time, and are emblematic of Japan's commitment to a potentially rapid expansion of floating wind power. Compare this to the rest of the world. The only installed utility-scale projects are Statoil's 2.3MW Hywind prototype off Norway (2009) and US start-up Principle Power's 2MW WindFloat off Portugal (2011). Both companies aim to build multi-turbine arrays using the same or similar technology, but these projects are still at the planning stage.

In floating offshore technology, the collaborative approach, with multi-industry R&D efforts and government-backed initiatives, should give Japanese developers, constructors, manufacturers and supply-chain participants a valuable industry-leading position. This could be exported to other potential markets, assuming that the technology can be proven and the cost of energy driven down. Opportunities in Japanese offshore wind may also open up for non-domestic companies, despite the market currently being something of a closed shop, with the major projects being built by consortia of the country's various industrial giants.

As we shall see, Mitsubishi Heavy Industries has merged its offshore business with Vestas, and the joint venture is likely to deploy the Danish turbine maker's 8MW V164 in Japanese waters, rather than the SeaAngel.

The Japanese floating wind projects are expensive, with a cost of energy of around 200–300 yen per kWh (about €1.40–2.10/US$1.95–2.90), compared to the most competitive European fixed-foundation installations at €0.11–0.12/kWh (US$0.15–0.16). However, it must be remembered that the pioneering Fukushima Forward and Goto projects were fast-tracked. Big savings are there to be had as the sector matures: it is an open secret in Japanese wind-power circles that the Fukushima consortium is looking to halve its cost of energy (CoE) for the project's second phase.

Much remains to be done in terms of regulation and in terms of engineering for Japan's offshore wind industry to really take off. However, a top Mitsubishi Heavy Industries executive confided to *Recharge* 'There is no other solution than offshore wind.'

Siemens cleans up

Before we look at which companies are in a position to win the leadership in offshore wind-turbine manufacturing, we need to look at how the contenders got to where they are at present.

According to Henrik Stiesdal, Siemens carried out the Bonus acquisition in 2004, as the Danish company was seen as one of the leading players in terms of quality and reliability, and because of its unique offshore experience. As well as building the first wind project at Vindeby, in 2003 Bonus had built what was then the world's largest project, the 166MW Nysted project with 72 turbines each rated at 2.3MW. In 2004, Siemens scaled up its turbine

Figure 5.2 Siemens' new direct-drive 6MW machine at the Gunfleet Sands offshore wind farm (Siemens/Dong)

offering with a 3.6MW model – the SWT-3.6–107 – that was to prove the world's most popular offshore turbine and become for many financiers the only 'bankable' machine in the market, particularly between 2010 and 2012. A 4MW upgrade of the machine was unveiled in early 2013.

Siemens has also invested in the installation process, buying 49 per cent of specialist vessel operator A2SEA in 2010 for DKK860m (US$160m). Siemens turbines have not been immune to quality glitches, such as a problem with corrosion protection that caused it to swap out the pitch bearings of its 3.6MW machines in 2010–11. However, the company proved that it had the operational strength to deal with the problem quickly and get turbines spinning again, and its reputation arguably emerged strengthened from the episode.

While the 3.6MW machine built a reputation for reliability, Stiesdal had become convinced that to make a permanent dent in both production and operations and maintenance costs, a new technology approach was required.

Siemens built its first direct-drive machine onshore in 2009, and began testing the prototype for its giant 6MW offshore turbine onshore in May 2011, with a second version with a larger 154m rotor diameter following in October 2012. In January 2013, Siemens installed two of the new 6MW turbines in an extension to DONG's Gunfleet Sands offshore wind farm in the UK. There are also plans to eventually build a 10MW machine.

Helped by the misfortunes that Vestas was suffering with its V90 machine, by 2009 Siemens had built its market share of offshore to 75 per cent, with partnerships with offshore-wind forerunners DONG Energy and E.ON playing a big role in its success. After a dip in 2010 when there was a surge in the number of Vestas turbines installed, Siemens' market share grew further and by 2013 it was 85 per cent. Even officials from Siemens say privately that this is 'not a healthy level' and they would like to see more competition. But that hasn't stopped the German company from hoovering up the lion's share of the most recently awarded offshore contracts.

Vestas gets burnt

One of the main reasons Siemens was able to become dominant in offshore is that the industry's biggest company, Vestas, has not been able so far to match its success onshore in the offshore space. Vestas constructed its first offshore wind farm, the 'Tunoe Knob' project in Denmark, as early as 1995, with ten V39 500kW turbines.

In 2001, it won the contract to supply 80 of its new V80 2MW turbines to construct Horns Rev, the first really big offshore wind farm in the North Sea. In December 2002, it finished installing the turbines ahead of schedule. However the turbines suffered a series of problems with transformers and generators over the following 18 months – badly insulated components were found to be the main problem – and all the 80 nacelles – plus the site's test nacelle – were taken back onshore for repairs in the summer of 2004.

Reports from the time describe a nightmarish experience. There was only one half-hour period during the whole 18-month period when all 80 machines were operating together. There were 75,000 maintenance trips – by helicopter – during the period, adding up to two per turbine per day. 'Experience is expensive, but also precious. Being the first large offshore project, Horns Rev must be a success', said Vestas' President and CEO Svend Sigaard. 'The project is important for Vestas' continued leadership in the offshore segment. It is my belief that Vestas will win the market in this segment. Even though it has been at a high premium, it puts Vestas and our suppliers into a unique position' (Vestas 2004).

The company had meanwhile won orders in the UK and the Netherlands. There is little doubt, however, that the issue caused lasting damage to the company's reputation in the offshore market, and worse was to come.

In early 2007, Vestas was forced to temporarily withdraw its V90-3.0MW offshore turbine from the market due to problems with the turbine's gearboxes, after 72 of a total of 96 V90-3.0MW turbines operating offshore developed major gearbox problems. It was only able to put it back on the market over a year later, on 1 May 2008. The V90 turbines that Vestas installed in Vattenfall's 90MW Kentish Flats project in 2005 were plagued with gearbox problems in their first four years of operations, until an upgraded version of the gearbox was installed in all the turbines in 2008 and 2009. The withdrawal from sale of the V90 delayed construction of Vattenfall's 300MW Thanet wind farm nearby and the £800–900m (US$1,350–1,500m) project, which became temporarily the world's largest when it was commissioned in 2010, also suffered initial gearbox problems after start-up.

Subsequently, however, V90s at Thanet and in the 180MW Robin Rigg wind farm in Scotland have performed well, according to the developers, with Robin Rigg developer E.ON reporting availability levels of over 98 per cent. The first units of Vestas' next offshore turbine, the V112-3.0, installed in E.ON's Kårehamn project in Sweden, have also been performing well.

However, although it was still picking up some orders for the V112, Vestas needed a big machine to get back in the game and compete for the big orders of the future. As we have seen, the company was forced to put back its initial plans to deploy and then mass-produce its 7MW V164 turbine in the UK. It then announced a slower-than-previous programme for building the V164 prototype and announced in 2012 that the turbine would now have an 8MW capacity instead of 7MW.

Building a prototype is one thing, but winning massive multi-billion-euro contracts from utilities for giant future wind farms in deeper-than-ever-before waters is another, and Vestas' executives had decided by early 2012 that they would need a heavyweight partner to really be able to compete.

REpower goes big

REpower, the German turbine manufacturer, was founded in 2001. In 2003, it launched the biggest offshore machine, the 5MW REpower 5M offshore wind turbine, at a time when the largest available model currently in the market was Siemens' 3.6MW machine. The company subsequently developed the 6M, with a 6.15MW rated power and deployed the first machine in 2009. As we saw in Chapter 4, the company was taken over in 2007 by India's Suzlon after a battle for control with French nuclear giant Areva, which has its own ambitions in offshore wind.

The first two 5M prototypes were deployed offshore in 2007 in the Beatrice demonstration wind farm in Scotland. Beatrice was a breakthrough project that deployed the turbines in deep water and supplied power to a nearby oil-rig. It involved a team of people, including Ronnie Bonnar and Dan Flynn, that would subsequently go on to play important roles in developing big offshore projects in Scotland. REpower went on to build Vattenfall's Ormonde wind farm in the UK, the Thornton Bank project off Belgium and formed part of the Alpha Ventus demonstrator in Germany.

Former REpower CEO Fritz Vahrenholt went on to become the CEO of RWE Innogy. The high point of REpower's commercial success so far has been the signing of a memorandum of understanding in early 2009 with the German utility for the supply of 250 of its 5M and 6M offshore turbines up to 2016. The turbines are expected to be deployed in the Innogy Nordsee 1

Figure 5.3 REpower's 5MW offshore turbine (REpower)

and Nordsee Ost wind farms in Germany, as well as in the Netherlands and the UK.

In 2010, REpower hired Siemens Wind CEO Andreas Nauen away from its German rival; the company looked to be on a roll. By 2012, REpower had a cumulative share of the offshore installations of 8 per cent, in a market still dominated by Siemens and Vestas. However, it was responsible for 19 per cent of installations that year, its best showing yet.

However, since then orders have slowed, as the company suffered the effects of the financial problems afflicting parent company Suzlon. It lost an order for the Gode 1 wind farm to Siemens after developer PNE sold the project to DONG. And at the EWEA Offshore 2013 conference in Frankfurt, CEO Nauen warned that the company would have to start laying off workers if new orders did not materialise soon. This was only a day after unveiling its biggest-ever turbine, the REpower 6.2M152, and a few weeks after announcing that the company was changing its name to Senvion, ahead of the rights to the REpower name expiring.

REpower has formidable strengths, having shown it can bring big multi-MW machines that perform reliability to market successfully, and ahead of its rivals. But there is a big question mark over its future as long as it belongs to Suzlon, and as long as Suzlon's finances remain under pressure. Arguably, in order to grow REpower needs to see a change in ownership if it is to resume growth and realise the potential of its technology.

Enter the industrials

The long-term strategic potential of offshore wind has attracted a number of heavyweight contenders to the arena, whose major selling point is widespread industrial experience and knowhow and balance sheets big enough to offer developers security in case of major failure in any offshore wind project. Among European companies, the new 'noisy neighbours' are France's Areva and Alstom, who we met in Chapter 2.

Areva lost a battle to take over REpower in 2007 and instead bought Germany's Multibrid, which had its own 5MW turbine design. It successfully demonstrated its M5000 5MW turbine in the Alpha Ventus project in Germany – dealing with a major third-party component failure in the process. With its HQ and production facilities in the German offshore-wind hotbed of Bremerhaven, it has since won a number of orders including the Global Tech 1 and Wikinger projects, and was also successful in the first French tender, winning the 500MW Saint-Brieuc zone. The company was also set to win its first UK order at the time of writing.

Areva hired highly respected industry veteran Julian Brown to run its UK operations in 2011, and the company continues to look at building manufacturing facilities in Scotland. However, its commitment to building a French turbine factory in Le Havre, and its existing Bremerhaven factory, give it considerable flexibility over the coming few years. Under pressure to compete with the big new machines planned by rivals, Areva announced in November 2013 that it would be bidding in the second French tender with a new 8MW design, and subsequently won 1GW of capacity with its consortium partners that included GDF Suez and EDPR.

Meanwhile Alstom deployed the offshore prototype of its 6MW Haliade machine in November off the coast of Belgium. Alstom won 1.5GW in the first French tender and committed itself to building two industrial hubs in France, while it has also identified manufacturing sites in the UK. Like Areva, having French manufacturing means that if it wins projects in the UK or elsewhere in the coming period it can supply these from France, while it waits to see the timing of the really big orders that will come from Round 3. It has hired former Vestas Offshore boss, Anders Soe Jensen, a highly effective salesman, to work with its VP Frederic Hendrick.

As well as Alstom and Areva, incumbent turbine suppliers see South Korean chaebol as a major threat. Samsung installed the prototype of its S-7.0–171 7MW turbine – currently the biggest in the world – at a test centre in Methil, northeast Scotland in the autumn of 2012, an extraordinary achievement for a company with a relatively short track record in the wind industry – and no experience at all in the offshore-wind game.

Although Samsung came late to the market, its deep pockets mean that it could potentially buy its way in, by extending project financing and guarantees. 'Ultimately it is going to be those with the biggest balance sheets that win', says wind-sector veteran and GL Garrad Hassan Chairman Andrew Garrad. He continues:

> It will be a case of having the vast suite of technological kit and finance. Samsung has a handful of land-based machines, but this is Samsung – if it wants to do something it will simply do it – it can't create a market, yet it certainly could take an equity share in a big offshore project to get it going.
> (Snieckus 2013)

Garrad adds, 'They may not have the cosy relationships of some of the others but they have come [to the UK] strategically to be part of the game. The impact of Samsung, if Samsung decides to play, will be very big indeed' (ibid.).

Although the first series-produced S-7.0–171s are most likely to be built far from Methil and installed at Samsung's own 84MW wind farm off the coast of South Korea's Jeju Island, one of the country's first commercial offshore facilities, the prize projects in Samsung's sights remain in UK waters. First large-scale sales – of at least 50 units – would allow the company to go ahead with plans to build a giant turbine factory in Methil at Energy Park Fife (EPF) – a renewables manufacturing and research complex set up by business development agency Scottish Enterprise (SE).

As well as Samsung, fellow chaebol Hyundai has developed a 5.5MW turbine and is looking at entering European markets, although its current plans are not clear. In addition, there are a number of Chinese companies, led by Sinovel, with offshore turbines in the water and plans for giant turbines of 10MW and above.

But perhaps even more significant than the Korean and Chinese players is the new joint venture that was announced in September 2013 by Mitsubishi Heavy Industries (MHI) and Vestas.

Vestas–Mitsubishi to challenge Siemens?

MHI had been developing its own offshore wind turbine, the 7MW Sea Angel using novel hydraulic technology, and had signed a framework agreement with UK utility SSE to both build and test a prototype and deploy the turbine in the developer's giant North Sea projects. This includes participation in the consortium that will develop the massive 9GW Dogger Bank Round 3 zone. But talks had begun with Vestas in early 2012 over a 'strategic cooperation' agreement. A year of sometimes-tense negotiations later and the deal was a reality. 'We have a clear strategic intent to become a global leader in the offshore wind industry', declared Vestas' buoyant Chief Executive, Anders Runevad, at the September announcement. The new company is in 'pole position' to win the race for market share, added MHI wind boss Jin Kato. The deal should propel both companies into the big time of the fast-growing offshore sector, creating a heavyweight competitor to undisputed market leader Siemens.

The agreement between MHI and Vestas revolves around the highly anticipated V164-8.0MW turbine. The development of the turbine will be transferred to the new company – along with Vestas' order book for its V112-3.0MW offshore turbines; its existing offshore service contracts; and about 300 employees. In return, MHI will inject €100m (US$135m) in cash and another €200m (US$275m) later, based on milestone achievements as the V164 is tested and brought to market. MHI will also bring considerable financial muscle and significant industrial expertise to the enterprise. 'Vestas needed a solid financial partner to compete with Siemens, Areva and Alstom on receiving future offshore turbine orders', says Patrik Setterberg, a senior analyst at Nordea, one of three financial companies advising Vestas on the deal. 'We argue that [the] Vestas/MHI JV will be a very strong alternative to Siemens if they successfully introduce the V164 turbine to market' (Backwell 2013e).

The talks, which were first confirmed by Vestas on 27 August 2012, took place – as we shall see in Chapter 7 – during the most difficult period in Vestas' history – a time when it faced a monumental financial hangover from rapid expansion, cost overruns, a cash-flow crunch and a major management upheaval. In late 2012, it seemed that creditors and banks were pushing the MHI deal as virtually the only way out for Vestas, while the disparity between the two companies' financial positions seemed to put the Danes at a huge disadvantage.

Industry sources say that the talks came close to being derailed over a number of issues – such as the value of Vestas' technology, who would control the JV and, according to some, MHI's attempts to gain access to Vestas' technology for its own onshore business, which has all but been destroyed by a series of patent disputes with GE.

At times, there was intense scepticism over whether negotiations were even continuing. Officials from Vestas always maintained their hope that the MHI talks would succeed, while simultaneously trying to assure investors that there was a 'plan B'.

Vestas' Chairman Bert Nordberg – no stranger to Japanese companies after running Sony Mobile Communications – told me in March 2013: 'I was CEO of a Swedish–Japanese JV so I learned the same patience as the Japanese. This is a marriage, not an engagement or a love affair. I am stubborn and I don't want to sign something that isn't good for Vestas.' He added that the delayed V164 would be developed with or without the MHI deal. 'We are not so weak that we have to force it', he said.

MHI's enormous fiscal resources are a huge asset for the new JV. With annual revenues of 2.8trn yen (US$207.6bn) in its 2013 fiscal year, MHI will almost certainly have the financial muscle to be able to extend guarantees to large-scale projects using the JV's turbines, giving it a potential edge over many of its rivals.

MHI also brings a wealth of expertise – in areas such as industrial outsourcing and aerospace – not to mention the capacity to build and supply offshore wind installation vessels. The Japanese giant is also well placed to leverage the growing political and industrial momentum behind offshore wind in Japan, as part of the key Fukushima project consortium.

In terms of the immediate financial impact on Vestas, the deal is 'marginal', according to Vestas. But the announcement had an immediate effect in dispelling doubts about the sustainability of Vestas' existing offshore business. The new JV's chief executive, current Vestas Asia Pacific and China President Jens Tommerup, said at the Copenhagen press conference: 'The JV will give us a strong business case to get some projects for the V112 before the V164 is ready, so we will also be stronger in the short term.' Later that same day, Vestas received confirmation of a sale for 43 V112s to Dutch utility Eneco – a relief after a long period without a single offshore order.

The new joint venture should also be able to take advantage of existing relationships with major potential customers. On Vestas' side, Danish utility Dong, the world's largest offshore-wind operator, is involved in early testing of the V164, and Tommerup says this will remain the case. Dong has been Siemens' biggest offshore customer, and has been eagerly waiting for Vestas to provide competition. Vestas has also worked closely offshore with German utility E.ON and Swedish utility Vattenfall.

Certainly the venture has got off to a good start. DONG announced that MHI–Vestas Wind is its preferred supplier for the Burbo Bank Extension project in the UK. The JV is also likely to sign several other contracts in the coming weeks and months.

Areva and Gamesa team up

In January 2014, Areva and Gamesa announced that they would be pooling their efforts in the offshore sphere and forming a new 50:50 joint venture. There is an inescapable logic behind the deal. Both companies are competing for a piece of the Northern European offshore sector and in recent months it has become clear that the development of the market would be considerably slower than had been expected when a large number of companies piled in a few years ago.

Both Gamesa and Areva had been looking at manufacturing sites in Scotland and, more importantly, both have a similar technology approach to their offshore turbines. Areva has a strong balance sheet, but it has sunk large amounts of cash into the offshore business with a so far meagre return, while analysts have long doubted Gamesa's ability to finance development of a large next-generation turbine, much less provide financial guarantees for multi-billion-euro offshore projects.

Some analysts have questioned whether the benefits of a deal were even for the two parties. Areva already had a limited number of 5MW turbines in the water, and is constructing two large German projects. It also has a contract for the 400MW Wikinger wind farm – also in Germany – and was a winner in the French offshore tender. It has a state-of-the-art nacelle factory in Bremerhaven and a blade factory in Stade and is building another industrial complex in Le Havre.

In contrast, at the time of writing, Gamesa had one 5MW prototype on a pier in the Canary Islands, and no orders. This does not tell the whole story, however. Gamesa has a vast experience in onshore wind and runs one of the biggest turbine fleets in the world. Its engineering knowhow is world-class, as can be seen in its groundbreaking 4.5MW and 5MW onshore turbine platforms. It has considerable capacity in component supply – the JV will enter into a preferred supplier deal with Gamesa for some key parts. And it has Iberdrola, the world's biggest wind-power developer, as its biggest shareholder.

Teams from Areva and Gamesa were set to compare the two companies' respective 5MW turbines to identify the best aspects and components of both and create an optimised machine. Engineers from Gamesa will also contribute to Areva's planned 8MW machine, the development of which is already well advanced. Analyst Robert Clover said:

> From a strategic and a cost perspective, there is no doubt that such a JV would strengthen the prospects for both companies, and will certainly make two Tier 2 players in the offshore market a more formidable presence, potentially enhancing their ability to jointly secure orders and importantly reduce costs to secure a profitable future for the combined entity.
>
> (*Recharge* 2014)

Who will be strongest?

The Vestas–MHI joint venture, on paper at least, should be in a strong position to challenge Siemens when it gets up and running in 2014. The company will have a strong, potentially game-changing product, financial and industrial strength, strong sales experience and almost unrivalled knowhow about wind energy. But it has a lot of catching up to do, and everything will depend on how the V164 performs. In a departure from the historically rather secret practices of the industry, Vestas will make data available directly to potential clients from the first prototype, as it pursues a strategy of early customer involvement in its product. The Areva–Vestas joint venture will also be in a good position to compete, as it will combine strong balance streets and the existing order pipeline of Areva, with the global wind expertise and experience of Gamesa.

Alstom, Areva and Samsung are all major industrial companies. While we have not looked at them in detail here, the Chinese companies are both ambitious, and can often bring their own significant financial resources to the table. With large-scale multi-year project portfolios likely to be the norm among developers, the panorama does not look encouraging for the independents.

On a wider level, it seems almost certain that further consolidation will be necessary in the offshore turbine market, with – in my view – only three or four being in a position to prosper in the European market. 'There are too many companies aspiring to be successful in the offshore market', says analyst Robert Clover. He adds

> Aside from costs of R&D and the balance sheet capability to honour any warranty issues required, there are too few unallocated MWs in the European market for all the present aspirants to reach sufficient economic annual manufacturing scale to warrant the new investments required.
>
> (personal communication)

6 After Copenhagen

A perfect storm for turbine manufacturers

Standing for several hours in sub-zero temperatures in the snow surrounded by riot police can be a sobering experience at the best of times. Journalists are no strangers to chaotic scenes – or queuing for that matter – but when I realised that members of the negotiating delegations attending the December 2009 Copenhagen climate talks were standing behind me in the same crowd I started to realise something had gone quite wrong with the organisation of the talks.

Even more sobering were the subsequent events inside the conference centre, when it became clear that the Danish hosts had underestimated not just the logistics of managing what had become a political and media circus, but the underlying difficulties of reaching any meaningful accord on a binding climate deal. The efforts of Denmark's Climate Minister Connie Hedegaard, German Chancellor Angela Merkel, the UK Energy and Climate Change Secretary Ed Milliband, Brazil's 'Lula' da Silva and France's Nicolas Sarkozy came up against the hard fact that neither China nor the US – the world's number one and number two carbon emitters, respectively – were interested in a deal. Christian Kjaer says:

> Leading up to Copenhagen, we were all quite optimistic. It was during Secretary of State Hillary Clinton's speech in the first week of Copenhagen that made us realise that the Obama administration was not on board and that the Danish Presidency of the COP had not been able to reconcile the positions.
>
> (personal communication)

The emerging countries – encouraged by China – putting their efforts into an extension of the inadequate Kyoto agreement; farcical interventions by supposed climate 'radical' (and major oil producer) Venezuela and its allies; NGOs missing the point of what was happening; and the heavy-handed tactics of Denmark's small but pit-bull-like security apparatus just added to the noise.

It did not help the process when the competent and experienced Danish Climate and Energy Minister Connie Hedegaard handed over the chairmanship to her newly appointed Prime Minister Lars Løkke Rasmussen for the second week of negotiations. Neither did it help that the Swedish EU Presidency negotiated a position that had not been agreed by the other EU member states. But COP 15 would have failed, even in the absence of these challenges. China and the United States did not come to Copenhagen with the intention of achieving a climate deal for the period after Kyoto and one has to wonder why Chinese Premier Wen Jiabao and US President Barak Obama showed up at all.

Even at the event itself, some wind-power companies were providing bullish forecasts of growth in the event of an agreement, something that was widely expected by the greater part of the industry. 'With a very strong support of the Copenhagen meeting, and a very clear road map from political leaders around the world, I think I can deliver by 2020 a $50bn size of the business', said Suzlon's Chairman Tanti (Stromsta 2009). Suzlon notched up sales of 260.8bn rupees (US$5.6bn) in the fiscal year ending 31 March 2009.

Tanti estimated that global turbine sales, which had averaged 28 per cent annual growth over the decade prior to Copenhagen, would jump to 35 per cent a year in the event of an agreement, a prediction that was far higher than the 22 per cent growth expected by GWEC.

Others were more wary. Faced with increasing signs that little progress had been made in negotiations ahead of the summit, Vestas' CEO Ditlev Engel said 'Let's see what happens in the coming days, I am still hopeful we can get things going' (Backwell 2009a). Engel highlighted the forthcoming presentation of EU countries' national action plans, and China's massive wind plans, as important factors that would keep the industry growing fast, even if an agreement at Copenhagen was not reached.

A few weeks later, industry leaders were putting a brave face on the failure of the talks, but it wouldn't be long before the wind sector was finding that wind directions had changed. After years of fast expansion, and a sellers' market where turbine companies could not ship enough turbines to keep customers supplied, the tables had turned.

Well before Copenhagen the effects of the 'sub-prime' financial crisis that hit the world headlines with the collapse of Lehman Brothers in September 2008, were making themselves felt. Bank lending for wind power projects had become extremely restricted, while companies found refinancing their debts increasingly difficult.

Bloomberg New Energy Finance estimated that by early 2009, investment in renewable energy was down 50 per cent from its peak a year earlier. 'The amount of capital available to finance projects shrivelled to nearly nothing, as liquidity problems made banks either stop lending for infrastructure altogether or demand tougher terms, including shorter payback conditions', said REN21's Virginia Sonntag-O'Brien (GWEC 2009: 5).

> Any money still left became very expensive, with financiers becoming more risk averse than usual … the amount of equity a project developer

had to provide to secure a loan increased dramatically in comparison with the pre-crisis days, when projects could be financed with as much as 90% debt.

(ibid.)

Making things worse, stock prices had dropped dramatically, preventing companies from raising money through new equity issues, while the recession, which resulted from the financial crisis, was slowing down power demand. Meanwhile, natural-gas prices were falling.

Significantly, the financial crisis was also hitting wind projects hard in the US, with the disappearance from the market of most 'tax equity' investors – typically big banks or corporations – who invest in projects to profit from federal tax credits. The financial crisis meant less profit and, hence, less need for tax credits. This meant that wind developers could effectively no longer make use of the Production Tax Credit (PTC), the main vehicle for US government support of the wind industry, even though it had been extended until the end of 2012.

Financing rebounded by the end of the year in great measure due to government stimulus programmes – in the US, for example, the government made it possible to convert the PTC tax credit into up-front investment grants – and multilaterals. Investment in clean energy was around US$145bn, only 6.5 per cent lower than the previous year's record figure (subsequent revisions suggest that in fact the dip in investment may have been exaggerated at the time).

And the wind industry continued to grow fast in 2009, by around 41.5 per cent compared to installations the year before and 32 per cent on a cumulative basis, or 38GW of new capacity, mainly due to one factor: China.

China's wind industry added a staggering – at least back then – 13.8GW of wind, more than doubling its capacity to 25.8GW, and overtaking Germany to become the world's second-largest wind-power market in cumulative terms behind the US. Bloomberg New Energy Finance reported 'extraordinary investment activity' in China in the last quarter of 2009. China accounted for 72 per cent of the US$4.7 billion raised through initial public offerings (IPOs) in the sector, and for one-third of the entire increase in globally added wind capacity.

For the international turbine manufacturers, however, China-centric growth was little relief, given that, as we have seen, they were effectively shut out of most of the growth and that prices in the Chinese market were at extremely low levels. In what looked like an ominous development for the mainly European incumbents, 2009 was the year that two Chinese companies, Sinovel and Goldwind, muscled their way into the top five global wind-turbine suppliers. And while companies like Vestas were able to post record financial results in 2009, orders for turbines – which are typically delivered 6–18 months later – were falling.

Europe slows while China growth skyrockets

In Europe, subprime had largely morphed into the Eurozone crisis in early 2010, with Greece making its first request for a bailout on 23 April, and Spain and Portugal suffering crises over the following months (Ireland had already begun the latest phase of its long-running financial crisis in late 2009). Other Eurozone economies were feeling differing degrees of pressure from net lenders such as Germany, to large but more bankable debtors such as the UK and Italy. Soon, governments in some of the fastest-growing wind markets were slamming the brakes on after years of spectacular growth, while others were thinking about how to reduce costs, even if this meant slowing down the transition to clean energy.

In 2010, the annual wind market decreased slightly for the first time in two decades, with the industry adding 38.3GW, or 0.5 per cent less than the previous year. Once again, China installed a new annual record of capacity – 18.9GW – and overtook the US as the world's biggest wind power in cumulative terms. Also for the first time, more new wind power was installed in developing and emerging economies than in traditional OECD wind markets.

In contrast to the continued ramp-up in Asia, the US wind market was only half the size of the previous year, at 5.1GW.

Europe's wind market was down by 10 per cent in annual terms at 9.91GW, with EU countries accounting for 9.29GW of the total. While the still small offshore market grew by 51 per cent, the onshore market decreased by 13 per cent. Eastern European countries were responsible for a large proportion of new installations, with Romania installing 448MW and Poland 382MW. Among the established wind powers, Spain was the largest market, adding 1.5GW, with Germany just behind, and France, the UK and Italy following. In January 2011, EWEA CEO Christian Kjaer commented:

> Remarkable growth in the onshore wind markets of Romania, Poland and Bulgaria could not make up for the decline in new onshore installations in Spain, Germany and the UK during 2010. These figures are a warning that we cannot take for granted the continued financing of renewable energy.
>
> (EWEA 2010)

In 2011, the pattern continued. The market rebounded somewhat, with 40.5GW of installations, 6 per cent higher than the year before. Once again, the majority of installations were outside the OECD and GWEC said the trend 'now seems firmly established' (GWEC 2011).

Global clean-energy investment reached a new record of US$260bn, with the key driver being a big rise in public-sector investment as part of government stimulus packages. State-owned development banks and agencies were among the top asset finance arrangers for clean-energy projects, including the US Federal Financing Bank (US$10.14bn), Brazil's BNDES (US$4.23bn), German development bank KfW, the Nordic Investment Bank, Danish EKF, the European Investment Bank, the World Bank and many others (see table GWEC 2011: 25).

China once again saw massive growth – 17.63GW – but this was less than in 2010. Analysts noted that the double- or triple-digit growth rates in capacity that the market had seen for nearly a decade were gone, and that the market was entering a phase of consolidation, with substantial excess manufacturing capacity. The decisive factor behind slowing growth in China was a lack of grid infrastructure to address the massive growth in the wind sector over the previous few years (see Chapter 3), and this would be a major factor holding back global rates of growth in 2011–13.

The US rebounded slightly, adding 6.81GW or 30 per cent more than installations in 2010. Momentum was once again building, in spite of – or because of – the uncertainty over whether PTC support would continue beyond the end of 2012. Significantly, the US industry was able to supply about 60 per cent of the manufacturing content for wind installations, compared to 25 per cent a few years before. This was a testament in part to the number of European companies that had set up new plants in the US in the last couple of years, including Vestas, Gamesa, Nordex, Acciona, REpower and Siemens. Capturing part of North American demand was key to the globalising strategies of these companies, but was clearly fuelling excess capacity throughout the industry.

Europe's installations were almost unchanged at 10.28GW, as the Eurozone crisis continued to stifle growth. Germany, the continent's most steady market, led with 2.08GW, followed by the UK (1.29GW of which 752MW was offshore), followed by the 2009 leader Spain, then Italy, France, Sweden and Romania.

A number of EU governments introduced measures to reduce support for renewable or began to look for measures to do so. As GWEC noted:

> On the one hand the EU has its 20-20-20 targets and on the other hand a budgetary crisis ... A troika comprised of the International Monetary Fund, the European Central Bank and the European Commission is watching over the budgets of Greece, Ireland, Portugal and Spain, and have in writing advised the Portuguese government to stop funding renewables via its budget ... This highlights the political risk involved in a political market.
>
> (GWEC 2011)

Spanish market's demise shakes the market

Spain's financial crisis and economic slump led to a rapid deterioration in political support for renewables, as the number one priority became reducing the fiscal deficit. While the solar PV and then wind sectors, as the fastest-growing energy sources, bore the brunt of political attacks, it is worth pointing out that Spain's energy demand had stopped growing and gone into decline, meaning that the country did not need new generation capacity of any sort.

Over the course of 2011 and 2012, the government introduced a series of measures, including a 'pre-register' that restricted the number of eligible projects that could still be built before the current regulatory scheme expired at the end of 2012. This was followed by a one-year moratorium on new projects – which in practice was extended indefinitely due to a lack of a new regulatory system for 2013 onwards, and a number of negotiated and unilaterally imposed changes to the feed-in tariff system. This meant that while there was still growth of more than 1GW in 2011 and 2012, these were effectively the last projects, and manufacturers were no longer clocking up sales from mid-2011 onwards.

Finally, in early 2013, the Spanish government imposed a new – highly unfavourable – regulatory scheme in the face of outright opposition from the industry.

The abrupt turnaround in the Spanish wind industry's fortunes had a major impact on the European industry. Spain had installed as much as 3.55GW in 2007 and 2.46GW in 2009. By 2011, the annual level had slumped to just over 1GW and by 2013 installations had practically dried up altogether.

The Spanish wind sector also had reverberations throughout the global industry. First, massive spare capacity was created from 2010 onwards, when the turbines for the last batch projects were produced. Spanish manufacturers Gamesa and Acciona suddenly found they had factories without markets, as did Vestas, Alstom and Enercon. Gamesa, in particular, had to practically reinvent itself to survive, as we shall see in a separate section.

Second, the giant Spanish utilities that had pioneered the global expansion of the wind industry, Iberdrola, Acciona and – Portuguese but Madrid-based – EDPR, found themselves fighting a rearguard action against retroactive changes that would impact their earnings from their already constructed projects, while also seeing their capacity to raise finance affected because of Spain's situation. In the case of Iberdrola, the company found itself having to face hard choices in its political lobbying as the government laid out its reform programme. Iberdrola was one of the two biggest Spanish power companies owed government money as part of the €25bn (US$35bn) so-called 'tariff deficit', a huge sum of money accumulated on utilities' balance sheets as a result of a long-running mismatch between accepted generation costs and artificially low consumer tariffs. With its interests in fossil fuel, nuclear and large hydro-generation, Iberdrola began adopting a high-profile stance as a critic of Spain's solar industry, and began to qualify its support for continuing expansion of wind power in Spain. From being recognised as the champion of Spanish wind energy, the company found itself in violent disagreement with the rest of the Spanish wind-energy sector over its role in the recent legislative changes.

With its balance sheet severely constrained, Iberdrola also had to make hard decisions about how it was to maintain its position as the world's biggest wind-power developer. The company had already begun to slow its rate of growth in 2011, mainly because of decreasing possibilities in signing power purchase agreements (PPAs) for new US projects.

The company had then attempted to shift growth into emerging European markets, and in particular Romania. But growth in Europe was harder to create than it had foreseen, because of delays in getting big projects off the ground and the growing climate of austerity. And finally, in 2012, the company began a programme of selling 'non-core' assets in some onshore wind markets in order to make an ambitious gamble on harnessing growth from offshore wind. In the future, the UK will be the primary destination for Iberdrola's wind development, along with some growth markets such as Latin America.

Overall, the Spanish crisis had a profound impact on the competitive landscape of the wind industry, as manufacturers lived or died on their ability to win market share outside Spain, along with an exodus of skilled wind executives seeking jobs elsewhere. The reasons why Spain's government stopped wind-power growth are understandable, but short-sighted. Power demand is no longer growing and the government has been in a situation of practical insolvency. However, the country continues to subsidise its coal industry as well as importing expensive fossil fuels, making a continued steady development of wind a good idea in the medium term. Managing a relative slowdown of the sector could have been handled differently in a way that recognised the strategic leadership position that Spanish companies had established – rivalling their Danish counterparts – in many areas.

Instead, the Spanish government has effectively taken a wrecking ball to one of its only high-tech growth industries, causing untold damage in terms of flight of expertise and capital. The fall-out has had an impact far beyond the borders of Spain: 'If it can happen in Spain, it can happen in any country is the logical conclusion made by wind energy financiers. They react by increasing the risk premium on financing everywhere', says Kjaer (personal communication).

US hits new peak, China slows

With European growth having levelled off, market growth depended more than ever on the US and China, together with some relatively small, but fast-growing emerging markets.

Continued uncertainty over the PTC and US government support for emissions reductions generally had the perverse effect of producing the biggest year ever for US wind installations in 2012, as developers rushed to install their projects before the threatened expiry of the scheme.

Installations were 13.2GW, with the US once again becoming the world's biggest annual market. Wind was the biggest source of new power capacity in the US – ahead of nuclear and thermal power – for the first time ever. As we have noted, the significance of this cannot be overestimated, given that gas prices were hovering at around US$3–4/MBTU throughout the period. Bloomberg New Energy Finance also noted that the majority of wind projects were built in states without renewable portfolio standards, meaning that utilities were buying wind power because they wanted to, not because they had to. On the other hand, massive growth in 2012 – with 8.4GW in the last quarter alone as the PTC deadline neared – was bound to set the scene for a big 'hangover' year in 2013, in a typical manifestation of the boom–bust cycle that short-term PTC programmes have created for more than a decade. US volatility has been one of the most difficult issues for wind-turbine manufacturers to handle, as illustrated by Vestas, which was laying off workers and considering plant closures in the US in a year of record wind-farm installations.

On the other side of the Pacific in China, the grid curtailment issue and the effects of too-fast expansion on inventories and company balances hit home hard in 2012, largely cancelling out the surge in US growth. China added 'only' 12.96GW, ceding the top place in annual installations to the US.

European growth picked up slightly, with 12.74GW of installations. Wind energy represented 26 per cent of all new EU power capacity, with wind meeting 7 per cent of Europe's power demand by the end of the year. However, as EWEA CEO, Christian Kjaer pointed out, the figures reflect 'orders made before the wave of political uncertainty that has swept across Europe since 2011. I expect this instability to be far more apparent in 2013 and 2014 installation levels' (EWEA 2013b).

In total, the market grew by 44.8GW, 10 per cent higher than the level of installations in 2011.

Wind shows its resilience

The wind industry faced significant setbacks in 2009–13, after almost a decade of outstripping the most optimistic projections. Overall, however, what is remarkable is not that the rate of growth of annual wind installations has slowed, or that there is a high level of volatility, but that wind power has been so resilient.

Underneath the yearly ups and downs, what is notable is that wind is competing, even when the most decisive factors – levels of government support and the comparative price of fossil fuels – are moving unfavourably. The real success story for wind, and the one that will ensure its long-term survival, is that it has become competitive with fossil fuels. Reaching the holy grail of grid parity is not clear-cut, due to the jumbled field of government support for renewables and fossil fuels, but it is clear that wind has reached some kind of competitive parity in many markets around the world.

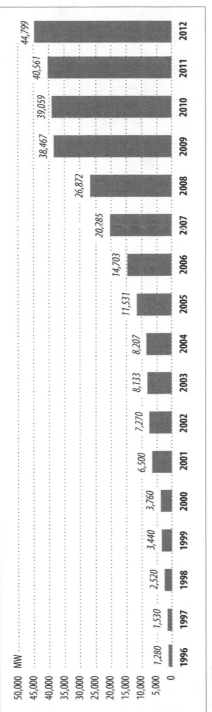

Figure 6.1 Global annual wind-power installations, 1996–2012 (GWEC)

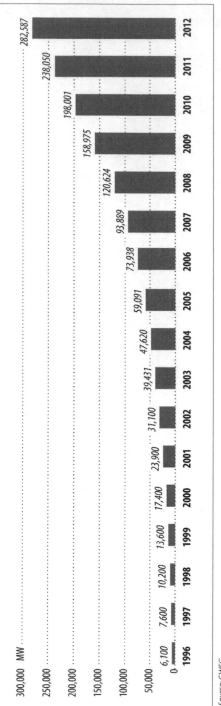

Figure 6.2 Global cumulative wind-power installations, 1996–2012 (GWEC)

As we have seen, in some markets, including Brazil, India, Australia and some parts of the US, wind is directly competitive. In other vast areas of the world, wind is also competitive now, as long as governments and regulators account in some way for the present and future cost of carbon emissions and pollution.

Ironically, the wind-energy industry is experiencing its hardest challenge at a time when the technology is the most cost-competetive it has ever been. The biggest challenge is no longer one of competitiveness – at least when it comes to onshore wind – the challenge is removing subsidies for all energy sources and creating a level playing field.

On the other hand, energy policy remains volatile, and to a great degree irrational, with short-term political expediency and market imperatives taking precedence over long-term outcomes, to a great degree due to the huge weight of the incumbent fossil-fuel complex. In my view, only a sustained and generalised fall in the cost of wind power can push wind to a tipping point where sustained rapid growth levels are assured.

At present, it looks like the industry is set for a major rebound in 2014, following a relatively slow year in 2013. The stock prices of wind-turbine manufacturers have recovered sharply from their lows of the summer of 2012.

But many of the issues that have affected turbine manufacturers in the most recent period are here to stay. We will examine how these issues could play out in our final chapter.

7   Turbine manufacturers in trouble

Vestas hits bottom

I flew over to Copenhagen on a bitterly cold day – 12 January 2012 – after being summoned along with other journalists for a major announcement on the restructuring of Vestas. Some changes had already been announced – or leaked – in the previous days, and it was clear that there was going to be blood on the carpet.

The backdrop to the restructuring was a series of worsening results, combined with two consecutive 'profit warnings' – the last in January – revealing big underestimates of losses which wiped out the company's 2011 profits and shocked investors.

'You have a company with little cost control, inadequate capital control and in some of their factories no operational control. It is business school 101 on how not to run a company', was the way one analyst, Martin Prozesky at Bernstein Research, described Vestas in 2012 (cited in Milne 2012).

After almost a decade of constant – even reckless – expansion, the seriousness of shrinking returns and slower than expected wind-market growth had finally hit home.

CEO Ditlev Engel announced in a sombre tone that 2,335 jobs would be cut, with the possibility of a further 1,600 being lost in the US. Whole units of the company were shut down and folded into each other, as the company took an axe to its costs and planned investments. The old geographical-based sales units were placed under a central global sales division.

Technology R&D was broken up, with parts of it going to the new Global Solutions and Services division – tasked with increasing revenues from outside the core area of turbine sales – and a new Turbines division headed by former Technology R&D Vice-President Anders Vedel. Global Solutions and Services also absorbed the existing Spare Parts and Repair unit. The Towers, Control Systems, Nacelles and Components and Blades units were merged into the new Manufacturing division. Wind, where Vestas had been pouring resources into its planned 7MW turbine and new production facilities, was also closed down as a separate division. The changes represented a significant streamlining of the management structure, with the number of executive managers being reduced from fourteen to six.

The casualties among Vestas' senior management, meanwhile, included offshore head, Anders-Søe Jensen; technology research and development (R&D) head, Finn Strøm Madsen; control systems head, Bjarne Ravn Sørensen; and head of communications, Peter Kruse.

Beyond the announcements, speculation centred around the fact that the man responsible – at least in formal terms – for getting Vestas' projected results so badly wrong and issuing the profit warnings – CFO Henrik Nørremark – was being promoted to an even more powerful position in the company.

The restructuring envisioned a new six-person management committee, with Nørremark as Chief Operating Officer and deputy CEO. The CFO position remained unfulfilled while the management board carried out an executive search for the right person. CEO Engel, whose leadership had been the subject of growing criticism, denied rumours that he had offered to resign at any point during the events of the previous months, saying 'The time to leave is not when there are challenges but when things are going well' (Backwell 2012a).

Most of the blame for the company's trouble seemed to be apportioned to the company's R&D department, which had underestimated costs on the development of Vestas' flagship V112 wind turbine and its Gridsteamer technology. Engel claimed that no blame can be attached to Nørremark for the delays in receiving revenues in the last quarter of 2011 and for a series of cost overruns, saying 'All the challenges that have surrounded Vestas in the last weeks basically come from one area: the implementation of new technology' (ibid.). When asked about the logic of promoting Nørremark into such a powerful position in the company, Engel said he could 'not think of a better person to be COO' (ibid.).

The elevation of Henrik Nørremark proved to be short-lived, and reportedly represented a desperate compromise by two men who were allegedly engaged in a long-running power struggle. In the struggle, Engel, a relative newcomer to Vestas, had the support of key institutional investors, while Nørremark was one of a group of company originals from Jutland, which also included Central Europe President Hans Jörn Rieks and Jan Pilgaard.

Insiders say that the board was extremely sceptical about the elevation of Nørremark – and Chairman Bend Carlsen had made disparaging remarks about the company's 'accountants' prior to the restructuring – and was wondering how they could sell such a move to the company's investors, who had just collectively lost their shirts following the profit warnings. Market reaction, a key auditor's report, and the realisation that more bad news was on its way pushed the board and Engel to make a new move.

On 7 February, Nørremark was unceremoniously dismissed. A statement from the company said: 'The board of directors of Vestas Wind Systems A/S has ... received a thorough briefing on the conditions which during the last months have led to profit warnings ... As a consequence of this, CFO and deputy CEO, Henrik Nørremark, resigns' (Backwell 2012b).

Ditlev Engel took over his responsibilities directly, joking that he was now 'CEO, CFO and CTO' (conference call with analysts and journalists on 8 February 2012), while the search for new executives continued. No explanations were given for Nørremark's dismissal at the time, and further bloodletting was taking place. Soe Jensen's replacement, Rieks, had been dismissed at the same time as Nørremark – once Engel had decided to move against Nørremark it was time to move against the whole group. His successor was former Ricardo and Areva executive Karl John, although his tenure was to prove short-lived. And on a wider level Vestas' programme of job cutting continued with white-collar and factory redundancies, aimed at reaching a level of 19,000 employees at the end of the year, from its 23,000-plus peak in 2010.

The backdrop to the management struggle going on at Vestas was a steady fall in the company's share price – Vestas shares fell 90 per cent between peaking in 2008 at 692 kroner (US$129) and the 58 kroner (US$11) mark level they had reached at the time of the restructuring. They were to fall further to a low of 23.25 kroner (US$4.3) on 20 November. In late 2012, up to 30 per cent of the company's shares were held by so-called 'short-sellers', 'investors' who had bought shares in the expectation that the price would fall further.

Along with successive waves of job losses in Denmark that the company announced in 2010 and 2011, Ditlev Engel had gone from poster child to villain in a short amount of time.

Relations with Vestas' syndicate of key bank lenders had also deteriorated, with the company announcing that it had not been able to fulfil its half-yearly financial covenant agreements when it reported its second-quarter financial results in July. Rumours began to circulate that the banks could force Vestas into a new share issue, something that could have been highly damaging to existing shareholders – and risky – in the prevailing negative climate.

Behind the scenes, Vestas' board was negotiating with Swedish telecoms executive Bert Nordberg to take over from long-running Chairman Carlsen, and it nominated him on 16 February. Nordberg became chief executive of Sony–Ericsson in 2009 and steered the sale of Eriksson's 50 per cent stake in the Sony–Ericsson mobile telephone joint venture to Sony in early 2012, becoming first CEO and then Chairman of Sony Mobile Communications. He was recommended by already serving board member Håkan Ericsson, who had worked with Nordberg at Ericsson in Silicon Valley. He carried out a wholesale realignment of Sony–Ericsson's product offering, and steered the venture through its takeover by Sony. But the company was also unable to regain significant market share under his leadership.

The appointment of someone with extensive experience of doing business in Japan and with the Japanese was almost certainly no accident.

Figure 7.1 Vestas' Chairman, Bert Nordberg (Vestas)

Nordberg took over the reins at the company's AGM on 29 March. On 24 August, Nordberg said that the company would seek a long-term strategic investor that could take a stake of up to 20 per cent in the company. On 28 August, Vestas announced that it was in talks with Japanese conglomerate MHI over a possible 'strategic cooperation' agreement. Although talks were known to centre on the capital expenditure demanding offshore wind sector, analysts speculated that the talks could lead to MHI taking a significant role in Vestas, or even taking over the company entirely, despite Nordberg saying, 'Vestas is Danish, and shall remain a Danish company listed on the Danish stock exchange. That is my ambition and that is the company's strategy. It is important for Vestas and also for Denmark' (Lee and Jensen 2012).

The news was received extremely positively by analysts and Vestas' lenders, who imagined a possible cash injection into Vestas that could head off the threat of any new share issue. Many of them probably didn't imagine how long negotiations with MHI would go on, and the highly secret talks became the focus for jittery investor sentiment over much of the next year. Several times, negotiations came close to the brink over a number of issues – such as the value of Vestas' technology, who would control the JV and, according to some, MHI's attempts to gain access to Vestas' technology for its own onshore business, which had all but been destroyed by a series of patent disputes with GE.

Nordberg told me in March 2012:

> I was CEO of a Swedish–Japanese JV so I learned the same patience as the Japanese. This is a marriage, not an engagement or a love affair. I am stubborn and I don't want to sign something that isn't good for Vestas.

He added that the delayed V164 would be developed with or without the MHI deal. 'We are not so weak that we have to force it', he said.

The talks were a cause of considerable anxiety, however. In February 2013 in Tokyo, Mitsubishi Wind boss Jin Kato seemed to deliberately ruffle feathers by telling me in Tokyo that the company was set to deliver 700 of its 7MW Sea Angel turbines – a clear rival to the V164 – to Scottish utility SSE for deployment in UK Round 3 projects. Kato and Vestas then boss Ditlev Engel seemed at pains to avoid contact during the same Tokyo event.

Aside from overseeing negotiations with MHI, Nordberg's main moves were continuing with job losses, trying to recruit high-level individuals to Vestas' depleted executive management and taking a broom to the company's internal problems. By making sure that the rest of the bad news lurking in the closet came out during Ditlev Engel's tenure – and I would argue that his reputation had suffered almost irreparable damage by this time – and that the most important job cuts were made quickly, Nordberg was ensuring that his successor would get a clear run and be able to bask in the glory of a good-news flow once the worst of the company's problems were behind it. Nordberg was also under pressure to make sure that any hint of financial impropriety had been dealt with in order to convince the intense due diligence that any deal with MHI would imply.

Nordberg appointed a fellow Swede Dag Andresen, a former Vattenfall executive, as CFO in April and Jean Marc Lechene, a veteran of the tyre manufacturer Michelin, as Chief Operating Officer at the end of June. The investor day held by Vestas in Aarhus on 3 October was all about 'cost out' – the process of running a rule over all of the company's operations in order to get production back to its original cost estimate – and other measures being taken to regain a positive margin.

Vestas attempted to balance the primarily conservative message of the investor day with the announcement that its planned 7MW V164 would now be an even larger 8MW. But there was no escaping the fact that there had been a clear reframing of the timescale for offshore within Vestas, and that it was doing everything possible to minimise and spread out capital investment on the new machine.

Vestas was able to meet its bank covenants and agree a new set of financial facilities in late November. It was clear that stability was returning as lenders could see its cost savings and cash-management programme advance, even though third-quarter numbers had been hair-raising, with free cash flow coming in at a negative €142m (negative US$181m), compared with a positive €276m (US$379m) in the same quarter of 2011.

On 7 November 2013, the company announced an 'intensification' of its cost-cutting programme that would see it shed an extra 2,000 jobs in 2013 and save an additional €150m (US$193m), bringing annual cost-savings to €400m (US$555m) since the end of 2011. The latest round of job cuts left it with a global staff of 16,000 by the end of 2013.

Chasing the Dragon

Vestas' restructuring programme had clearly put the company on the track towards breaking even in 2013 and a return to profitability in 2014.

Meanwhile, the talks with MHI continued, despite frequent rumours that the talks had ended without success, or that Vestas' lenders were trying to force a deal with the Japanese by the beginning of 2013.

However, the management battle within Vestas had still not played itself out, nor was the position of Ditlev Engel, the 'last man standing' of the company, by any means secure.

On 2 October 2012, Vestas announced that it had terminated Henrik Nørremark's severance package in September after learning that he had entered into 'two commercial agreements' in India on behalf of the company without informing other senior management, or the board, causing the company a loss of '€4m (US$5.17m), possibly up to €18m' (Backwell 2012c). Nørremark contested the claims and began legal proceedings over the issue.

At the time, Vestas did not go into details over what the Indian deals involved, and Nordberg said a 'complex investigation' was under way (ibid.). Officials made clear that they considered that the transactions were fraudulent and could lead to criminal charges being filed against Nørremark.

However, the following May I reported that the allegations involved a mysterious attempt to buy wind turbines in India from Chinese competitors. Nørremark is alleged to have set up a top-secret project called 'Project Dragon', by which Vestas' R&D department in India was to have set up a 100MW wind farm involving an unspecified number of Chinese machines.

Officials from Vestas told me that Vestas' Contract Review Committee, which approves all turbine contracts over 10MW, was not informed of the deal, and nor was Vestas' CEO Ditlev Engel. Apparently, Vestas did not take delivery of any turbines, nor was any wind farm constructed. CEO Ditlev Engel had handed over his computer and all his emails to a PWC forensics team, and although one Danish newspaper reported it had seen emails between Engel and Nørremark showing that Engel did know about the project, they did not publish them. Nordberg said in a statement that the inquiry by the company's auditors and lawyers 'proved that neither the board nor the group president and CEO have been involved in or have had any knowledge of the mentioned transactions' (Backwell 2013g).

In July, Danish newspapers reported that Denmark's Public Prosecutor for Serious Economic and International Crime (Fraud Squad) had formally started proceedings against Nørremark.

Also in July, Vestas filed two lawsuits against its former Indian joint-venture partner RRB at the High Court of New Delhi, after starting arbitration proceedings in the International Chamber of Commerce in June, in an attempt to recover €24m (US$32m). Group Senior Vice-President Morten Albæk said:

> Vestas is determined to reclaim the money which has been transferred to ECO RRB by instruction of Henrik Nørremark without authorisation, as well as the debt which was cancelled by Nørremark, inter alia, without there being any business-related explanation to it.
>
> (Stromsta 2013c)

Indian company RRB Energy and its London-born owner Rakesh Bakshi were Vestas' original joint-venture partners in India. The two companies parted ways in 2006 when Vestas sold its 49 per cent stake in Vestas RRB after acquiring NEG Micon, which had its own presence in India. However, they maintained close ties, with Vestas agreeing to focus on 750kW-plus turbines and RRB concentrating on smaller machines. Vestas agreed to continue to assist RRB through a technological cooperation agreement. Danish papers highlighted a private dinner that took place between Vestas' CEO and Bakshi in the summer of 2008, although this was three years before the disputed payments were made. Sources in India point out, however, that Henrik Nørremark was very close to Bakshi during his time in India.

The true nature of Project Dragon has remained a mystery. Some of the questions which were still outstanding at the time of writing are:

* Did Vestas really try to buy Chinese turbines to analyse them, or as Vestas officials believe, was 'Dragon' a simple case of fraud?
* Were any turbines actually delivered to Vestas/RRB and if so, who was the supplier?
* Who else in the company management knew about 'Dragon'?
* Did RRB Energy use Vestas' money to buy Chinese turbines for its own purposes (the company was working on a new 1.8MW design at the time)?

The worst is over?

By mid-2013, Vestas seemed to have weathered the storm. The company had announced orders of over 2GW as of early July, with the second-quarter intake the strongest quarterly figure since 2011. Patrik Setterberg, senior analyst at Nordea, said that the company's strengthening order intake was 'not a coincidence, but rather a sign that Vestas' customers have greater trust in the financial position of the company' (Backwell 2013h).

Analysts MAKE pointed out that Vestas was in the best position to take advantage of a small global increase in orders – there was an increase of 48 per cent in volume compared to the same period in 2012. 'On the data we have seen for the first half, Vestas is performing the most consistently across the three regions', said Research Director Robert Clover (Backwell 2013i).

Investment bank HSBC noted that Vestas, along with other turbine manu-facturers, had taken 'bold steps' to restructure its business and that the cost of restructuring measures had already been booked in 2012.

Meanwhile, Vestas' share price had recovered by 270 per cent in July from its low point in summer 2012, helped by investment banks like HSBC and Nordea putting 'buy' or 'overweight' recommendations on the stock. By early December 2013, the shares were back up to around 150 kroner (US$28), almost 400 per cent higher than a year earlier.

'A future without surprises'

The recovery in Vestas' financial fortunes was not enough to save Ditlev Engel, however. On 21 August the company announced that Engel was leaving and would be replaced by Anders Runevad, another Swede, and another former executive from Ericsson and Sony–Ericsson.

While thanking Engel for his work over the previous eight years, Chairman Nordberg said, 'We feel that after eight and a half years we are looking for a tweak of the culture and more focus on profitability, and a future without surprises' (Milne 2013). The remarks were polite but cut-ting, given that in his last few years as CEO Engel had presided over a man-agement war, four profit warnings and at least two big investor lawsuits against the company.

Vestas' shares rose 5 per cent on the news of the new CEO. The mes-sage from Runevad was clear: management would be focused on the sober tasks of efficiency and cost management, in close liaison with the company's major shareholders.

The move was a classic in ruthless management. Vestas' board had effect-ively waited until all the bad news was out of the way, and the bulk of unpopular job cuts had been made, before 'terminating the terminator', in the words of one former Vestas official who had lost his job back in 2012. The *Financial Times* cited a person 'close to' Bert Nordberg as saying 'For us it was important that we weren't in the most vulnerable situation. He was looking for a time when the situation was more stable so you could give more time to the new chief executive to settle in' (Milne 2013).

The more generous analysts pointed out that Engel had been better at building up Vestas than in managing it when times got tougher. Others criti-cised Engel for pushing ahead with big global expansion plans at a time when the global economy was about to go into recession and renewable-energy support schemes were threatened, while its investments in China soon became a liability as the country's own manufacturers took over the market. 'A different management skill set is required to run a business for growth than to restructure a business, and the two skill sets are often not present in the same individuals', says analyst Robert Clover (personal communication), who followed Engel's fortunes closely first at HSBC then at MAKE Consulting and Recharge Insight. It was also reported that institutional investors that had lost big on Vestas' shares during the crisis period would not come back while Engel was in charge.

Runevad is set now to preside over a period of positive developments. But whatever Engel got wrong, the wind industry lost one its most charismatic figures when he resigned, and one who had been determined to propel wind into the big leagues of international energy business and politics.

Mitsubishi comes through

A few weeks later, on 27 September, Vestas and MHI announced the creation of an offshore joint venture at closely coordinated early-morning press conferences in Copenhagen and Tokyo. It was over a year after talks were first announced. Officials from the two companies said that they expected the new joint venture to challenge Siemens for the top spot in the offshore wind business (see Chapter 5).

Back to profit

When Vestas released its full 2013 results in early February 2014, it seemed to have delivered on the main areas of its two-year turnaround plan.

The company pulled in full-year revenue of €6.084bn (US$8.44bn) in 2013 and its earnings before interest and tax (EBIT), before special items, was €211m (US$293m), with a free cash flow of €1.009bn (US$1.4bn). The EBIT figures were 44 per cent better than analysts' consensus expectations.

Vestas said that 'the higher-than-expected revenue and EBIT were primarily driven by a smooth execution in terms of installation and transfer of risk combined with favourable weather conditions in December' (Lee 2014b). For 2014, it expects revenue to amount to a minimum of €6bn (US$8.3bn) with an EBIT margin before special items of at least 5 per cent and a minimum free cash flow of €300m (US$416m).

The figures showed how deep Vestas' cost-cutting measures had bitten. Annualised fixed capacity costs had been lowered by €484m (US$671m) compared to the fourth quarter of 2011. Net investments had shrunk by more than €500m (US$690m) to €239m (US331$m) since 2011 and working capital had been lowered to negative €596m (US$826m).

Vestas ended 2013 with orders of 5.96GW, up from 2012's 3.74GW. It delivered 4.86GW against 6.04GW. 'A double-digit EBIT margin in the fourth quarter and a free cash-flow generation of more than €1bn in 2013 are major achievements for Vestas', said CEO Anders Runevad (Lee 2014b).

What next for Vestas?

The restructuring and the exit of most of the Vestas 'historicals', the MHI deal and Engel's departure marked the end of an era for the company.

It's too soon to predict Vestas' future fortunes. Although the management from now is likely to be both more efficient and focused on financial management, the structural problems that the company faces remain: increased competition from big industrial companies like Siemens and GE, a slower market in Europe, difficulties in competing in China and a highly cyclical US market that could leave it with big overcapacity.

On the other hand the company has formidable strengths: a geographical diversity second to none that allows it to gain from any small increase in global market growth; a huge resource of knowledge and experience of wind energy; strong products and the advantages that having the world's largest existing portfolio of turbines – 57GW or 20 per cent of the world's total – brings; and the backing of government in the world's most wind-friendly country.

For the foreseeable future at least, as long as the wind industry exists, Vestas will exist in some form. However, the question of ownership and scale remains. Vestas' investor base is fragmented and, unlike the joint venture with MHI, its balance sheet is weak compared to companies such as Siemens or GE.

A look at the histories of Mr Nordberg and his associates shows a history of selling on companies to bigger investors – at the time of writing he was on the committee tasked with selling failing mobile telephone company Blackberry. Vestas is no longer Danish-run, and there is no guarantee that it will remain a Danish company in the future at all. Although there is no indication of this at present, MHI could be back for a further bite – and this time of Vestas as a whole – at some point in the future.

Rivals feel the pain

Of course, Vestas was not the only company that suffered during 2011–12. Hard times hit companies in different ways and, as we have noted, there was some considerable shelter for companies that were part of larger industrial groups.

Life was undoubtedly the most difficult for publicly quoted wind-power specialists. One – Clipper, the second-largest turbine producer in the US – teetered on the edge of bankruptcy, was bought out by a larger group and then shut down.

Gamesa adapts to Spanish meltdown

Another, Gamesa, had to make an even tougher transition than Vestas. Gamesa had been one of the biggest beneficiaries of both the Spanish wind boom and – linked to this – the expansion of the world's largest wind developer, Iberdrola. As we have seen, Iberdrola is also the company's biggest shareholder with a stake that has oscillated around 15–20 per cent of the total. The two companies had successive framework agreements that meant that up to 50 per cent of Gamesa's business was coming from orders from Iberdrola.

Iberdrola's annual installations reached a high point of 1.78GW in 2010, but installations slowed down after that. The Spanish energy crisis stopped development there and it slowed down its breakneck expansion in the US in 2011 as it found it harder to sign new PPAs. Attempts to make up the shortfall by speeding up in Europe – notably in Romania and elsewhere in Eastern Europe – came up against transmission and regulatory obstacles, while its balance sheet – like many other European utilities – came under increasing pressure.

The company's three-year plan for 2012–14 foresaw investments of €2.6bn (US$3.35bn) in renewables, compared to €1.9bn (US$2.6bn) annually between 2009 and 2011. The company said it would add 1.45GW of new wind power during the period, compared to over 1GW per year in the three previous years. It also announced a major sale of 'non-core' wind assets and a big move into offshore wind.

Gamesa was a major casualty of the slowdown. The huge contracts it expected in Romania and elsewhere failed to materialise, and it realised it could not take Iberdrola for granted when it came to contracts. Gamesa was behind the game in offshore, and building new offshore turbine platforms was a capital-hungry affair at a time when its cash flow and debt figures were moving in the wrong direction.

As we have seen, Gamesa made aggressive moves to enter the Indian and Chinese markets, while trying to maintain sales in China through pursuing close relationships with the big utilities and by the end of the first quarter of 2011, 100 per cent of the company's sales came from outside Spain. The company moved to downsize much of its Spanish manufacturing. It also reduced capital expenditure on its planned 5MW and 7MW offshore turbine programme, moving a planned prototype of the first of its new machines from offshore in Virginia to a quay in the Canary Islands.

Success in emerging markets was not enough to stop Gamesa going into the red, and it reported its first quarterly loss in over a decade in March 2012, while its shares had fallen around 75 per cent over the previous year. Nor was it enough to save the job of Executive Chairman Jorge Calvet, who was replaced in May the same year, reportedly after pressure from Iberdrola. The new Executive Chairman was Ignacio Martin, yet another veteran of the auto industry – a former CEO of automobile components Cie

Automotive SA (CIE) – who had been recruited to rationalise costs in the wind industry. In October 2012, Gamesa announced a major restructuring that would reduce Gamesa's work force by 20 per cent – a loss of 1,800 jobs – and align the company's operations with a new lower level of annual demand, by shutting down five manufacturing facilities, reducing capacity by 2GW.

Financial analysts were impressed with the speed and aggressiveness with which Gamesa faced its problems, and compared the company favourably with Vestas' difficulties in restructuring.

By early 2013, the worst was over and, unlike Vestas, Gamesa was back in the black by the first quarter and for the first nine months it reported a net profit of €31m (US$43m), turning around a €67m (US$93) loss at the same stage in 2012. Share prices had been recovering steadily from their 2012 low point. Martin told *Recharge* in Chicago in May:

> Previously Gamesa was focused on growth and expansion. Now the main focus is on performance. If the wind market is growing, that's excellent. But if there's no growth, then we still need to perform. We've reduced our break-even point ... we're launching new products and we're continuing with our geographical expansion so that we're never dependent on any specific market.
>
> (Stromsta 2013a)

The company's flagship 5MW upgrade to its revolutionary 4.5MW onshore turbine looked as if it was beginning to gain traction, with its first large-scale orders in Finland. It began energy production from its 5MW off-shore prototype in the Canary Islands in October.

The company's challenges are far from over, however. Continued success in the markets where Gamesa has done well in recent years is far from guaranteed. As we have seen, India's market slowed down in 2012 and 2013, along with China. In Brazil, Gamesa has made an impressive grab for market share, but the new, more challenging FINAME regulations mean it will struggle to maintain profitable margins. And in the US, it is subject to the same stop–go market that has made managing a global wind business so difficult for its peers.

In offshore, it is arriving late to a market that already has a number of experienced incumbents and aggressive newcomers. It remains to be seen whether it can make an impact with a 5MW turbine, while plans for a 7MW machine are a long way from fruition. The long-standing relationship with Iberdrola is no guarantee of sales, as can be seen in the Spanish utility's decision to use Areva turbines in its 400MW Wikinger Project, and work with Siemens on its first UK project (which it is building with Danish utility DONG Energy). Like Vestas, Gamesa has a formidable accumulated knowledge of the wind market and highly regarded products. But like its larger Danish cousin its balance sheet cannot compare to the diversified industrial companies and it is 100 per cent dependent on the ups and downs of the global wind market.

Suzlon struggles with debt overhang

As we have seen, Indian turbine manufacturer Suzlon went through a meteoric expansion in the years leading up to the Copenhagen summit at the end of 2009, and – taking into account its REpower subsidiary – became the world's fourth-largest wind-turbine manufacturer.

The company's Chairman Tulsi Tanti had based his calculations on continued rapid growth of the global wind market. And while Suzlon was a leader in the onshore sector, offshore was taking much longer to get off the ground than Tanti had expected. A blade-cracking issue affected Suzlon's S88 wind turbine in 2008, forcing the company to pay around US$100m for a large-scale remediation programme.

The blade issue helped drag Suzlon into the red in 2009, and the losses continued in 2010, while the debt it had incurred to carry out and complete the REpower acquisition was becoming an ever more serious burden. Meanwhile, in China, Suzlon – in common with other non-Chinese players – found it was being squeezed out of the market by extremely low prices and margins, after having invested heavily in production facilities there. Suzlon's problems led to officials at rival Gamesa floating the idea of a merger or take-over – although it is not clear how seriously this was actually pursued, and Tanti was quick to stress that the company was not for sale.

Although the company had some successes, both in emerging markets like Brazil, and in offshore, its financial problems worsened throughout 2011–12. Although Suzlon's home market of India slowed as a whole in 2012, due to the phasing-out of Accelerated Depreciation, the most notable development is that Suzlon was not able to execute the orders that it had in its backlog because of restrictions on working capital. This led to Suzlon losing its leading position in the market in terms of annual installations for the first time to former Enercon subsidiary, Wind World (India). In a statement, Suzlon said 'Fiscal Year 2012–2013 for the Suzlon Group was focused on liability management, which resulted in a constrained business performance and lower than normal volumes, despite a strong order backlog' (Backwell 2013j).

Although Suzlon completed its 'squeeze out' of REpower in late 2011, attempts to use the profitable REpower division and its ability to raise finance to alleviate the parent company's balance sheet were unsuccessful, as REpower's German banks maintained a strict policy of ring-fencing. Despite announcing on several occasions that a more thorough integration of Suzlon and REpower was on the cards, observers were left somewhat bemused by the fact that no really major moves emerged, despite the prevailing overcapacity in the wind-turbine industry. Rumours of a 'reverse takeover' where the subsidiary would take over the parent company also turned out to be without substance, or the idea proved to be impractical.

Throughout the period, rumours circulated that Suzlon would be forced to sell off REpower to pay its Indian bank lenders and bondholders. In March 2012, a French engineering giant denied rumours that it was preparing a €1.5bn (US$2bn) bid to buy REpower. Officials from Suzlon also denied reports that it was trying to sell REpower, calling rumours 'completely speculative', although I was told by reliable sources that the company had opened a data room in Germany for potential investors through an investment bank.

The backdrop to the rumours was the looming issue of US$569m of payments on foreign-currency convertible bonds (FCCBs) later that year. In October 2012, Suzlon carried out India's biggest FCCB default – on US$221m of bonds – playing hardball with investors that refused to meet new terms, while continuing to work on a new deal with its syndicate of bank lenders.

According to investment bank HSBC 'Post the default on FCC repayments in October 2012 the company's operations (ex REpower) have come to a standstill, given the freeze on its bank accounts' (Charanjit Singh, HSBC, 24 June 2013, private correspondence). At the same time, Suzlon began a US$400m asset-sale programme, selling off – among other things – its China manufacturing business, which it had quietly shut down some time before.

In January 2013, it announced it had agreed a US$1.8bn package with a consortium of 19 banks to restructure its domestic debts under India's Corporate Debt Restructuring (CDR) mechanism, in a move that it said would help 'normalise' its business. The deal included a two-year moratorium on principal and term-debt interest payments, a 3 per cent reduction in interest rates and a six-month moratorium on working capital interest.

CFO Kirti Vagadia said the move was a 'major step forward in our efforts to achieve a sustainable capital structure' (Lee 2013b). At the time of writing, negotiations were continuing with bondholders. The CDR deal will probably allow Suzlon to reclaim its top spot in India in 2013, and the company should benefit from a speed-up in its home market from 2014 onwards, but Suzlon continues to be very cash-constrained. Sources in the company say that Indian lenders are pushing the company to sell off most of its non-Indian assets. Certainly, the evidence from once promising markets like Brazil suggests that Suzlon is carrying out a wholesale retreat from global markets.

Although officials deny that they are under any pressure from their bank lenders to sell REpower – renamed Senvion in November 2013 – the question of what to do with the German subsidiary will continue to be asked. In a nutshell, REpower could have the potential to be more successful than its parent company and, according to most analysts, the heavily indebted Suzlon is acting as a drag on the German company's prospects. At the same time, Suzlon is unable to fully integrate REpower without damaging or destroying it, while the debt overhang from the REpower acquisition continues to limit the parent company's ability to operate successfully.

IHS-EER Research Director Eduard Sala de Vedruna says, 'REpower is vital to keeping the Suzlon Group afloat, but Suzlon's debt position inhibits REpower's capacity to finance new product developments or provision off-shore warranties.' He warns that Suzlon needs to clarify 'the German vendor's future strategy inside or outside the Suzlon Group to avoid further value destruction' (*Recharge* 2013).

Officials from Suzlon continue to say that REpower, the 'crown jewels' of the company, is not for sale. Privately, however, officials say that if the right buyer came along with perhaps €1.5bn (US$2bn) to spare, the company would be open to discussion. Although my view is that something has to give in the Suzlon–REpower relationship, and soon, the most common refrain among Suzlon and REpower managers for now is 'only Tulsi Tanti knows' what is going to happen.

Clipper – goodbye to the 'other' US OEM

The history of US turbine manufacturer Clipper, which enjoyed big success at the end of the last decade, can be traced back to the origins of the US wind industry and Zond. Clipper was formed in 2001 by Zond Energy founder Jim Dehlsen and his son, a few years after the Zond business had been sold first to Enron and then to GE.

Clipper expanded fast in the latter part of the decade, producing its 2.5MW Liberty turbine. Its US installations peaked at over 605MW in 2009, while it was also designing and producing the giant 10MW Britannia offshore turbine in the UK – mainly with UK government money – that was meant to spearhead the technology drive needed to develop the giant Round 3 zones.

While its technology was well regarded, and the Liberty was in some ways ahead of the game in a market dominated by GE's 1.5MW turbines, the company had several major weaknesses. First, like other OEMs it was subject to the US stop–go cycle and, unlike its rivals, it did not have a global manufacturing and market presence to compensate. Second, Clipper's growth was inordinately dependent on one company: oil major BP, which in the latter part of the decade was making a major commitment to wind energy. BP had backed Clipper's IPO in 2005, and the two companies were working together on the 5GW Titan project in South Dakota – one of the largest planned wind-farm complexes in the world.

Clipper was hit hard by a blade-cracking problem in 2007, which largely deprived it of customers outside of BP and hit its balance sheet hard, costing the company US$330m in a remediation programme to replace the blades on all its turbines. Then BP began to apply the brakes to its wind business growth and orders slumped, effectively sealing the fate of Clipper.

Clipper was teetering on the edge of bankruptcy by late 2009 after a loss of US$109m in the first half of the year, and put itself up for sale in September. Industrial conglomerate UTC – which produces among other things Sikorsky helicopters – completed an investment of US$207m to take a 49.5 per cent stake in Clipper in January 2011, and gained full control in for an additional US$223m in December that year.

At the time, Clipper CEO Douglas Pertz said that the sale to UTC was

> a transformational transaction for Clipper, bringing substantial capital from a strategic investor who is one of the world's leading industrial technology companies. Our relationship with UTC will enable Clipper to access UTC's support and expertise in areas of manufacturing, product quality and other industrial processes, while providing Clipper with equity financing to deliver our longer-term strategic goals.
>
> (Backwell 2009b)

Officials at UTC subsidiary, Pratt & Whitney Power Systems, said they were looking at plans to develop a 'next generation' wind turbine.

In the event, UTC took a long hard look at Clipper and the wind-turbine industry and decided that it would be one guest too many at the party. UTC moved to cancel Britannia in August 2011. In August 2012, it sold Clipper to private equity group Platinum Equity for an unreported amount. At the time of writing, Platinum seemed focused mainly on trying to realise some value from maintaining Clipper's existing turbine fleet, and the company's production facilities were wound down in late 2012.

'UTC bought Clipper under certain market assumptions', says DaPrato, 'and most of them proved wrong. While the total investment was pocket change for a company of that size, UTC didn't want to wait five years to see if it could possibly pay off', he adds (Kessler 2012).

'We all make mistakes', Chief Financial Officer Greg Hayes said in March of UTC's brief flirtation with Clipper, after putting it up for sale (ibid.).

8 Challenges for the wind-turbine industry

The period 2010–13 was a traumatic one for many in the wind-power manufacturing industry. And although a new period of growth is now ahead of us, the issues that emerged most strongly during the period are unlikely to go away anytime soon. I want to look at some of the key challenges facing the wind sector by identifying some of the unresolved contradictions at work in the industry.

Political and market volatility versus globalised manufacturing

The first challenge is market volatility. The growth of the wind industry has been a story of spectacular, followed by steady, growth. But the global figures mask a series of stop–start patterns that can be extremely abrupt. As Alstom's wind boss Alfonso Faubel puts it when comparing the wind industry to the global automobile manufacturing industry, 'One of the biggest challenges we have is lack of consistent volumes across the board. Automotive has consistent volumes' (personal communication).

Growth rates in the more mature markets have slowed, while new countries are continually coming into the market. Given wind's complex interaction with incumbent power-generation complexes, carbon and the energy-subsidy landscape, in the majority of cases the key variable until now has been the prevailing level of political support, which in turn is linked to economic cycles and the level of fiscal solvency of different countries.

Within the EU, the longest-standing and most constant market, there has been steady growth in onshore wind in Germany and Denmark, where political support has been strongest. Steady but slower growth has also been the norm in France and the UK. Other countries like Spain, Portugal and Greece have seen precipitous growth followed by a virtual standstill in the aftermath of the Eurozone crisis, with Italy also taking measures to slow down market development. Newer markets, such as Romania and Poland in Eastern Europe, have seen extremely fast ramp-ups followed by sharp slowdowns as governments have realised the short-term impact on the public finances, and the political environment has changed.

Meanwhile, in the US, the creation of the PTC scheme as the preferred tool for supporting the country's renewables sector and the absence of any federal portfolio requirement has led to the most extreme boom–bust cycle in the industry. Changes in the regulatory framework also caused an abrupt fall in growth rates in India, when the accelerated-depreciation and GBI incentive schemes were allowed to expire.

While political volatility is without a doubt the main issue, it is not the only one. In China, the issue of the lack of available grid connections, followed by widespread curtailment, have been the primary factors in slowing down growth after the spectacular 2008–11 period. The fiasco around delays in building grid connections also virtually halted growth in Germany's offshore sector in 2012, and played a key role in curtailing massive planned investments in Romania. Brazil is currently threatened by similar issues, with around 1.2GW of wind farms currently not connected to the grid. On a wider level, a continued lack of interconnections between European countries has created bottlenecks that have prevented the wind industries in some countries from maintaining growth by exporting power.

Another issue is the availability of private finance. The subprime/Eurozone crises in 2008–11 restricted financing for wind projects in the US and Europe. The Chinese government subsequently moved to restrict funding by state-owned banks, also affecting developers.

And there are more issues, such as the impact of fossil-fuel prices both on power spot markets and the availability and cost of capital for projects – the shale gas phenomenon has had the most marked impact in recent times, by breaking the historic relationship between US oil and natural-gas prices and contributing to bring natural-gas prices down to a range of around US\$3–4/MBTU (with a low point of US\$2/MBTU) – as well as the underlying question of the evolution of power demands of different countries.

Bouts of volatility can effectively bankrupt wind-turbine companies. Firms that have invested heavily in new manufacturing plants, sales and service operations to capture fast growth can find themselves scrambling to mothball capacity and relocate or lay off staff a few years later. Think, for example, of Spain's Gamesa, that effectively found itself sitting on factories with nothing to produce from 2011 onwards, or companies like Vestas, whose shiny new US factory complex was virtually idle for much of 2012, once the rush of orders that had been supplied that year were delivered.

Wind-turbine companies have taken different approaches to the problem. The most common, and that which has been pursued over most of the last decade by heavyweight OEMs such as Vestas, Siemens, Suzlon and Gamesa, has been to deal with volatility by increasing its level of globalisation. Former Vestas CEO Ditlev Engel summed up the philosophy when he told me in 2012, 'We need more globalisation and not less' (personal interview). According to this strategy, being in more markets provides a hedge against constant changes in the rate of growth in different countries. Producing and selling in one market means that if it fails, it is effectively game over.

Producing in a wide variety of markets means that you can capture new growth, while having a diverse enough flow of revenues that one stalling market will not kill you. More local production also means less exposure to volatile currency exchange rates, and means that you are more able to influence the local business environment.

Globalisation is also widely considered key to the wind industry's ambition to bring about a sustained reduction in turbine production costs, and therefore in the cost of energy from wind power. Lowering turbine costs requires both a globalised supply chain for components and the kind of scale that only large global manufacturing players can bring about. According to the proponents of this argument – and we will look at this in more detail further on – the creation of the necessary scale requires industry consolidation and the growth of a small number of big players, as the wind industry follows the path of the automobile sector and other major industries.

In practice, however, the globalisation strategy has been hard for companies to follow. This is particularly the case for the 'specialist' wind-turbine companies that are not linked to wider industrial engineering groups. Restricted balance sheets and the pressure to produce consistently positive EBIT has meant that most companies have found it hard to sustain the level of investments necessary for anything like true globalisation.

The company that has come closest in terms of the number of manufacturing facilities and global sales network is Vestas. To this day, the Danish company is often the best placed to take advantage of newly emerging wind markets, as well as uniquely winning a major share of orders across the three key areas of Europe, the US and Asia.

However, the company has also suffered more than most from overcapacity and the cost of mothballing and restarting manufacturing operations, and it does not have the same financial muscle as its conglomerate competitors. It has found it particularly hard to cope with the US stop–go cycle. The company invested fully US$1bn in a complex of manufacturing plants in the state of Colorado, only to have to mothball them months after completion. After having laid off and compensated a large number of staff, it found itself having to rehire many of them to deal with the rush of deliveries it faced in 2012 – only to once again have to virtually mothball factories during the PTC hangover lull of 2013. Vestas has had numerous other experiences of setting up production facilities – the Isle of Wight in the UK, Tasmania in Australia, Spain and Italy spring to mind – only to shut them down because of a 'lack of market' a few years later.

A number of Vestas' competitors – Gamesa, Nordex, Acciona and Alstom to name a few – have also found setting up in the US to be difficult, while China has been an even tougher proposition. Suzlon tried hard to compete on terms with its Chinese rivals, with Chairman Tulsi Tanti once famously telling non-Chinese companies to 'stop complaining and start competing' (Backwell 2010b) in China. However, by 2012, both Suzlon and its

REpower subsidiary were out. Joint ventures set up by Acciona and GE ended in failure, while Nordex tested the water and decided that selling in the Chinese market was a bridge too far. By 2013, only Vestas, Gamesa, GE and Siemens – through its Shanghai Electric joint venture – were still attempting to stay the course and treat China as a sales market and not just a place to source components.

Being able to compete everywhere is not a given. India has proved to be an extremely hard place to do business, due to the particularities of the local business environment. On the whole, only those with deep local experience and a willingness to get involved in wind-farm development and land deals have been able to command a sizeable slice of the market. Vestas found that it lost market share quickly when it moved to a supply-only model, and among non-Indian companies only Gamesa – precisely because of its willingness to adapt to local conditions – has been able to significantly expand.

Brazil may also turn out to be an expensive trap for some, with at least eight wind-turbine manufacturers – including Vestas, GE, Gamesa, Alstom, Enercon, Acciona, Suzlon and Siemens – piling in to one of the world's most competitive markets in price terms, and one that is unlikely to add up to more than 3GW of installations annually.

Betting on niches

A second option is to pursue a niche strategy, concentrating on one or more key markets where your company feels it has an advantage.

The most obvious example is German manufacture, Enercon, which has consistently won the lion's share – 54 per cent in 2012 according to IHS-EER – of its home market with an expensive but top-quality product and an unparalleled service model. Outside Germany, Enercon has been successful in being an early entrant into some other markets, usually in response to government tendering, including Brazil, Canada and Portugal, but it has stayed out of major markets such as the US and China.

While some of its international policies seem to have been due as much to management eccentricities as to strategy, it is hard to argue with Enercon's business model. Fully 38 per cent of its sales come from Germany, where it is able to sell at premium prices; and Germany's onshore market has been the most stable, steadily growing wind market in Europe, and indeed the world. Enercon has not faced the brushes with bankruptcy that competitors such as Vestas or Gamesa have faced. Its technology and products are widely admired and popular among its customers.

On the other hand, Enercon is exposed. During the recent political debates around Germany's Energiewende, it was interesting to note both the sudden willingness of traditionally shy Enercon company officials to engage with the media and policy debates – and their nervousness. As one competitor pointed out, 'They have more to lose from things going wrong here than any of us' (private communication).

Other companies following a niche strategy are Nordex, which under the new leadership of CEO Jurgen Zezschy has pulled out of competition in China and the US, while being successful in building its market share in its home market of Germany – and in spotting niche opportunities in places like Turkey, Pakistan, Sweden and Uruguay. Referring to the company's down-sizing of its international manufacturing footprint, CEO Dr Jurgen Zeschky says in his letter to shareholders in the company's 2012 report:

> I am confident that with the leaner structures which we have imple-mented we will be able to make full use of the advantages which we have in the market place. This is because we have always been success-ful as a mid-size company. Customers do not expect Nordex to be a big corporation but, rather, a flexible engineering partner that is able to respond quickly and understands all aspects of their business.
>
> (Nordex 2012: 5)

Another, very different example is France's Areva. As we have seen, the French-owned nuclear giant fought an unsuccessful battle with India's Suzlon to take over German turbine company REpower in 2007. It then took over the smaller German designer and manufacturer Multibrid and built up a formid-able manufacturing and marketing operation focused solely on the offshore space. Areva considers that with the big balance sheet and the experience in large-scale power plants of its parent company it is in an advantageous pos-ition to win a share of the big orders that utilities will have to place for off-shore turbines in the coming years. Onshore wind with its greater scale, but far smaller average order size, would be merely a distraction.

Of course the divide between 'fully globalised' and 'niche players' is far from clear-cut, and can be better understood as two opposite ends of a curve.

As an example, Tier 1 player GE – either number one or two in the glo-bal ranking in 2012 according to the leading analysts – pursues a highly selective approach to the markets it enters, while making sure it maintains leadership in its home market of the US, helped as we shall see by synergies within the GE group. GE has an extremely lean manufacturing footprint that relies on complex logistics to deliver projects. Main components for the giant Fantanele wind farm in Romania, for example, were produced in Brazil where GE has plants.

Spanish-owned turbine manufacturer Acciona Windpower has largely scaled back its production to its plants in Spain, avoiding overreach and keeping its core facilities viable. It has also mainly geared its manufactur-ing and logistics to supply the operations of the wind-farm development activities of its parent company, Acciona Energia. However, being a niche manufacturer has not stopped Acciona Windpower from starting up new facilities to supply third party customers where it sees a market opportunity, with the company recently announcing major investments to meet stringent local content requirements in Brazil.

And it is worth looking at Germany's Siemens, which has historically derived a large part of its total wind-turbine sales from offshore, even as it pursues a strategy of trying to increase its market share in the main onshore wind-turbine markets, propelling itself into third place in the global rankings in 2012.

Cross-industry players

We should consider at this point that dealing with geographical and generalised volatility – i.e. the varying rates of growth across the global sector – in the wind industry is to a great degree a different proposition for the 'non-specialist' wind-turbine companies, i.e. those that form part of much larger industrial companies. The main factor is the size of company balance sheets. In simple terms, wind-turbine manufacturing losses are not likely to bankrupt parent companies. A loss that could cripple Vestas (revenues €6.084bn (US$8bn)) is likely to be less significant for Siemens (revenues €78bn (US$108bn)), Alstom (€20bn(US$28bn)) or GE (US$147bn).

At the same time, most of the big industrials possess a natural 'hedge' because of their other activities in the wider energy sector. If a company like GE, Siemens or Mitsubishi has a bad year for its wind-turbine equipment division, it is likely to have had a good year for its division that produces turbines for gas or coal plants.

Being diversified also allows companies to shift personnel to high-activity areas and make use of synergies across manufacturing assets. Famously, GE was able to ride the rollercoaster of US turbine deliveries during the record year of 2012 by assembling wind turbines in its gas-turbine manufacturing plants, allowing it to take first place globally. It was then able to shift personnel back to their usual duties a year later when the wind market entered a lull. Others had no such advantages on either the upward or downward part of the cycle, and faced expensive ramp ups and then layoffs of personnel.

It is worth remembering, however, that although they have bigger balance sheets, parent companies have shown that they will try to impose strict financial discipline on their wind-turbine divisions to avoid any deterioration of company-wide EBIT targets. This means, even though bigger balance sheets are in play, wind-turbine divisions in practice arguably face similar restrictions on their activities as the specialists. On the other hand, the existence of the parent company means that clients are unlikely to be scared off by doubts about the overall solvency of the turbine supplier, in contrast to what has happened on occasions over the last two years with some of the 'specialists'.

An example is Vestas. Nordea analyst Patrik Setterberg says that the strengthening of the company's order book in early 2013, once the company had weathered the worst of its financial troubles 'is not a coincidence, but rather a sign that Vestas' customers have greater trust in the financial position of the company'. He adds 'We believe that some customers reduced counterparty risk in 2012, negatively influencing order intake' (*Recharge Magazine*, 3 July 2013).

Overcapacity and consolidation

The expansion of OEM's global manufacturing operations and the entry of new manufacturers into the market led to a situation of over 100 per cent overcapacity in the market by 2012, which inevitably led to lower turbine prices and a prolonged squeeze on turbine manufacturers' margins.

According to Robert Clover, who has headed research into wind-sector companies for both investment bank HSBC and sector analysts MAKE, there was around 85GW of manufacturing capacity for turbine nacelles compared to a demand of around 40GW. He notes that a significant part of the excess capacity built during the period was in China, where, as we have seen, at one point in 2011 there were over 70 OEMS in the market.

Clover identifies a number of factors behind the build-up of excess capacity in 2008–10. The main causes were the high returns in the wind industry, which attracted large amounts of capital, and the simultaneous onset of an economic slow-down, slower growth in power demand and regulatory pushback. 'The wind industry before the crash expected compound annual growth rates of something like 20–25% over the 2008–2013 period', he says, 'and this ended up being more like 6.5%.' He adds that 'The excess capacity really built up in 2008–2010 as investments planned before the crash came online.' Indeed 2008–10 saw new companies continuing to enter the market and build new plants.

'We went very fast from a situation of excess demand to excess supply; from high prices to low prices; from long delivery times to manufacturers chasing customers – even those smaller ones that they had dismissed in the race to scale in the upturn', says Christian Kjaer (personal communication).

Looking forward, Clover says he does not expect annual installations to reach more than 66GW even by 2020, meaning the turbine manufacturing industry needs to contract. Even before the crash, observers of the wind industry had been predicting a phase of consolidation in the wind-turbine market, to follow expected trends in global manufacturing industry development.

From 2011, it seemed that the writing was on the wall. While Vestas, Gamesa and Suzlon all seemed to be heading for insolvency, there were persistent rumours of large-scale industry takeovers. These included various Chinese or Japanese players taking over Vestas, Gamesa merging with Suzlon, Alstom buying REpower, Siemens buying Nordex or Enercon and so on.

None, so far, has turned out to be true, although MHI and Vestas have, as we have seen, formed a joint venture and Gamesa and Areva are looking to follow suit. Wind-turbine manufacturing companies were certainly affordable in 2012 – and still are – but potential buyers are perhaps understandably reluctant to take over an OEM in a sector where global supply outstrips demand by almost 100 per cent.

On one level, most agree that consolidation is an inevitable step for wind. As Suzlon's Chairman Tulsi Tanti says, 'Every mature industry has a small group of leaders, the ones that claim most of the market share', adding that the wind sector will 'emerge with its own big five' (Tanti 2012). Tanti points out that the basic drivers for the process are that wind is both 'extremely capital-intensive' and 'hugely cost-competitive' (ibid.). Yet large-scale consolidation at an OEM level has yet to occur, despite the hardships, low margins and massive overcapacity of the last couple of years. Why not?

First of all, those that could feasibly buy their competitors have decided not to. Companies like GE, Siemens and MHI have decided not to. When I asked GE's Vice-President for renewables Vic Abate in 2012 why GE didn't simply buy out its cash-strapped competitors, he replied by asking why GE should buy excess capacity in a situation of 'rampant overcapacity', adding there was a danger of simply acquiring out-of-date industrial technology. 'When you think of consolidation you think of a buyer and a seller,' he says, 'but I ask myself "what can't I do organically?"' (Backwell 2013f).[1]

When I asked Alfonso Faubel, Alstom's Senior VP for wind power, about large companies being able to simply swallow up their smaller competitors, he said, 'You need to preserve shareholder value. Meaning you can't just spread around the world with new factories; it doesn't work. You need to use cash very carefully' (personal communication).

Both officials said that consolidation would continue, but that the dominant manner in which this would occur would be through bankruptcies of those companies whose balance sheets would be inadequate for the newly challenging conditions of the market.

Companies have also cited incompatibilities between their technology strategies and those of potential takeover targets. This has particularly been the case for offshore, where producers of big new direct-drive turbines have seen existing geared products as redundant for their growth plans.

The reluctance to buy out competitors reflects a strong streak of cautiousness in the head offices of the big industrial firms. As I have mentioned, company boards do not want to countenance anything that will erode margins. And reluctance to buy surely reflects underlying uncertainties about the future market size, which is ultimately a factor of wider energy and climate politics and policies. In other circumstances, surely the companies which have the balance sheets would simply buy to take out competitors and ensure that they would take a bigger market share when the industry starts a new growth phase?

Along with established firms' cautious policies in a situation of overcapacity, another factor which could have played a significant role in preventing large-scale international consolidation from taking place was the slump in Chinese turbine manufacturers' fortunes from 2011 onwards. Chinese manufacturers that had seen several years of stellar growth and almost unlimited access to new cash had been eyeing major European acquisition targets as part of their 'go global' strategies. But 2011 saw their financial conditions tighten as the government imposed new conditions on state banks' lending, while the effects of major overcapacity in the Chinese market became clear soon afterwards. In the event, Goldwind did acquire the German turbine designer and manufacturer, Vensys. But any acquisition of a Tier 1 supplier will have to wait for a future wave of Chinese expansion.

Meanwhile, although the big headlines on consolidation have yet to materialise, consolidation has been occurring among smaller companies and component manufacturers. Clipper Windpower, the only other large US wind power apart from GE, was close to bankruptcy in 2010. As we saw in Chapter 7, it was subsequently wholly bought by technology and industrial giant United Technology Corporation (UTC) in December 2010. After a period of evaluating Clipper's business, UTC effectively pulled the plug on the business and sold it on to private equity fund Platinum in August 2012. Among small turbine manufacturers to hit bankruptcy are Germany's Fuhrlander – once a significant second-tier player – German offshore specialist BARD and Finland's WinWind.

There has also been a process of consolidation among key component suppliers, with a combination of large-scale acquisitions, for example in gearboxes, and bankruptcies.

But the question of what level of spare capacity the industry needs and can afford to maintain, and how many companies will survive, is still very much pending.

In my view, widespread consolidation will indeed occur – finally – at some stage over the next five years. Several major manufacturers continue to be constrained by weak balance sheets, some suffering from a debt overhang which they will find difficult to shake off.

Economic recovery, the availability of cheap money struggling to find a competitive return and an increase in turbine demand is likely to make consolidation more, and not less likely to occur, as increasing confidence is likely to allow the wind divisions of the major industrial players to persuade cautious head offices that a new wave of investment in wind is a good bet.

A long-predicted increase in Asian activity in mergers and acquisitions is finally likely to become a reality. On the one hand, major Japanese and Korean corporations like MHI, Marubeni and Samsung are building up their knowledge and expertise of European wind markets, particularly in the areas of capital-hungry offshore wind-farm and power-transmission projects. MHI's formation of a joint venture with Vestas and its buying into a number of German and UK offshore transmission hubs is likely to be a sign of bigger things to come. Marubeni has bought into offshore wind installation through its acquisition of Sea Jacks, as well as directly into UK Round 3 offshore projects. Korea's Samsung has built a 7MW offshore turbine in Scotland, but will find it will almost certainly have to buy its way into the development market in some way if it is to carve out a major presence.

On the other hand, Chinese companies are likely to take up the government's favoured 'go global' strategy with renewed strength, after the hiatus of 2012–13. Renewed growth in the Chinese market will give the leading companies financial oxygen, but competition will remain intense, and foreign markets will begin to present attractive acquisition targets, particularly as the renminbi continues to rise in value, making exports less competitive. A small group of leading companies such as Goldwind continue to build their international expertise, including developing and selling projects in non-Chinese markets, raising finance and certifying their technologies for global markets.

It is impossible to predict how many turbine manufacturers are likely to be in the market by the end of the decade. Most officials from major companies predict that there will be around five to ten, compared to the dozens of players operating today. In my view, this is a reasonable estimate, if we imagine perhaps three or four major European-owned players, one in the US and perhaps another three or four based in Asia. However, as industry observers have pointed out, predictions of large-scale consolidation have so far proved to be premature.

Vertical integration versus the networked industry

Related to market growth and industry capacity is the issue of what kind of business model wind-turbine companies should follow.

For years, the model for the big tier-1 wind-turbine companies was 'vertical integration', a management system that emphasises control through different levels of the supply chain. Vertical integration is used for example in the oil industry, to describe companies that combine both 'upstream' oil exploration and 'downstream' oil refining and marketing.

In the wind industry, the major wind turbine companies have controlled manufacturing of everything from steel hubs through to gearboxes, electrical components such as generators and control systems through to towers, as well as the blades and nacelles which most industry officials agree make up the core of the business.

Vertical integration became the norm during the boom years of the wind industry in 2005–09, when turbine manufacturers were fighting to ensure that components were available to them during periods of high demand. Leading manufacturers carried out a series of acquisitions of supply-chain companies or acquired the necessary technology to make the components in-house. 'Today there is a clear reversal of that trend', says JP Makinen, CEO of leading power-electronic supplier The Switch. 'Turbine companies are concentrating on their core competencies, while outsourcing production, because they have come to understand that this is the only way to increase supply-chain flexibility', he adds (personal communication).

Makinen advocates a 'networked' approach, with open collaboration between companies being the new industry norm. 'What's new in the networked business environment is that all parties involved are sitting around the same table, working towards the same result', he says (ibid.), adding that the key to successful collaboration is building trust among all parties.

Makinen's appeal for openness is far from being the accepted norm among the big turbine producers, some of whom are almost obsessively secretive about their technologies. Tensions still arise continuously between turbine OEMs and their component suppliers – the most common cause of high-profile turbine failure continues to be non-performing bought-in components – as both sides attempt to maximise returns, and this is likely to be the case for the foreseeable future.

It is clear, however, that the wind-turbine industry is changing under conditions of spare capacity, and that companies are reassessing what they see as core business, while looking to squeeze more value from their operations by shedding operations where they do not enjoy any specialised competencies.

Leading turbine manufacturer Vestas has carried out an extensive outsourcing programme, including selling off a major tower factory in Denmark and its metal-casting manufacturing, and outsourcing logistics, warehousing and transport operations. The company has also outsourced key components for its flagship V164 8MW turbine. It has also begun to use its tower production facility in Colorado, USA to produce towers for third parties, to minimise spare capacity and bring in extra revenues.

India's Suzlon sold off gearbox manufacturer Hansen back in 2011, and has since moved to sell off other manufacturing units such as castings.

There is of course no clear-cut divide between companies taking a vertically integrated approach and those taking a more flexible approach to the supply chain. US turbine producer GE, for example, buys its blades from Brazilian supplier Tecsis through long-term supply deals and others have a mix of in-house and third-party-produced blades. Up-and-coming turbine producer Alstom made an early decision not to get into the blade business and has sourced the blades for its giant Haliade offshore turbine from blade specialist LM Windpower.

The debate about outsourcing is part of a wider discussion about what model for industrialisation the wind industry should follow. As we have noted, auto manufacturing has been seen as a key model by company executives, and over the last couple of years, companies have hired a number of managers from the auto industry.

In June 2012, Vestas appointed Jean-Marc Lechêne, former director of Michelin's heavyweight tyres unit in Europe, as its chief operating officer, with responsibility for manufacturing and sourcing. At the same time, Spain's Gamesa appointed Xabier Etxeberria to the new role of 'business chief executive officer'. Etxeberria comes from the automotive arm of engineering group GKN, where he spent ten years in the Driveline division. Etxeberria was employed by Gamesa's recently appointed Executive Chairman Ignacio Martín, who himself worked at GKN and more recently was Executive Vice-Chairman of car industry component and sub-assembly supplier CIE Automotive.

French industrial giant Alstom was even further ahead of the curve. It appointed Alfonso Faubel – who had held senior positions in US-owned auto component and technology supplier Delphi – as head of its wind business back in 2009.

It is worth pointing out that the level of mechanisation within the sector remains relatively low. Most nacelle assembly plants involve a large warehouse with groups of workers moving around from workstation to workstation fixing the different sets of components into place, before the finished product is hauled out and loaded onto its transport.

Alstom's Senior VP for wind Alfonso Faubel, an auto-industry veteran and a strong advocate for modernising production systems, describes the industry in transition from being a 'Mom-and-Pop shop kind of industry, to a truly industrial set up where automation, where building quality, where all best practices from supply-chain management were being utilised' (personal communication).

An interesting case is that of Siemens, which, like Alstom, has extensive company-wide experience it can bring to bear from large-scale industrial manufacturing. Siemens was the first wind-turbine manufacturer to introduce a true moving production line in Brande in 2010, using a flow production concept designed by Dürr Consulting. Siemens says the system has reduced assembly time per nacelle to 19 hours from 36 hours, and the number of workstations from 18 to 8. Dürr says the digital three-dimensional factory layout it completed in April 2009 'forms the essential basis for all new Siemens wind-turbine assembly lines all over the world, especially in the Asian and American regions' (Backwell 2010c) and allows Siemens to globalise its nacelle-manufacturing operations and set up factories wherever there is demand.

CTO and founder Henrik Stiesdal says that he looks closely at the truck-manufacturing business, which he says provides a more realistic point of comparison with wind than the consumer automobile industry in terms of the number of units produced. He predicts a growing level of automation in production, although he says that the industry's likely maximum scale implies that full robotisation is not on the cards.

Siemens' strategy is based on the automotive industry's practice of mass-produced modular components, while wind turbines are 'bundled' into product platforms. 'We can reduce production and logistics cost by standardising and modularising components within our product platforms', says Stiesdal, describing the strategy as 'a major step towards achieving our goal of making wind power independent from subsidies' (personal communication).

For Stiesdal, the main barrier to automation in nacelle manufacturing is the amount of work involved per unit – Siemens assembles a 2.3MW nacelle in less than 250 hours. 'The possible benefit from automation is much more likely to be at the level of the component suppliers', he says. 'For blades it's different and this is where we are focusing the automation effort' (ibid.).

True to his profile as one of the wind industry's most consistently independent figures, Stiesdal says that what companies actually end up physically producing will continue to be debated, and even puts forward the idea that a company like Siemens could still be a turbine manufacturer without actually producing nacelles. 'You can actually imagine companies being more "knowledge based" rather than owning industrial assets', he says, pointing to rival turbine manufacturer GE, which outsources everything apart from nacelles. However, he adds 'There is no obvious best strategy – Enercon is equally successful with a very high degree of vertical integration' (ibid.).

While the general thrust of big industry players is towards automation, we have seen there are powerful forces that are have put limits on the intensity of capital and processes like automation.

The most important is geographical volatility. As we have seen, the wind market consists of a series of rising and stabilising or falling national markets, making it difficult for wind-turbine manufacturers to plan and implement a global manufacturing strategy. This is exacerbated by the common use of local content requirements, as national government's try to 'capture' wind-power investment and employment. Rising markets such as Brazil, Turkey and Canada all have some kind of local content requirements, as have previous growth stars such as Spain and Portugal – whether these are through government tenders, or through restrictions on financing as is the case in Brazil.

In the most basic terms, the demand for local manufacturing – as socially and politically necessary as it may be – can prevent components for wind turbines being produced at the lowest possible cost and mitigate companies' attempts to introduce real scale advantages due to the limited size of individual markets.

Wind-turbine companies, which were obliged to set up local manufacturing to address the fast-growing Brazilian market due to restrictions on BNDES financing for developers (see Chapter 4), aimed to set up local manufacturing in the cheapest and most cost-effective way. However, doubts over the total size of the market in the future meant that they were unlikely to invest in state-of-the-art mechanised plants.

An example of the effects this environment can have is Vestas' decision in 2012 to change from a system of blade manufacture it had introduced for its V112 turbine in Europe – whose purpose had been mass standardised production – to a new design based on the tried and tested method of individually crafted blades. The reason: the need for a 'capex light' industrial footprint which allows the company to quickly – and cheaply – set up operations in new markets.

Clearly, Vestas' decision made sense for the company, in the context of its offensive to reduce capital expenditure costs. From an industry-wide perspective, however, a promising move towards industrial scale that should have eventually led to a sustained reduction in cost was halted in favour of an earlier technique that is more costly in terms of man-hours per unit.

Again, the tension between global manufacturing scale and local production is not an out-and-out dichotomy. The Switch, for example, is championing a system of 'model factories' that can be set up cheaply and quickly in any market, and which it argues also embody best manufacturing practices.

There are clearly still many issues to be worked out over the coming period. One can hope that a combination of some degree of consolidation and an upsurge in market growth will allow companies to give some answers to these questions. As I argued in one of my columns in 2012, wind-turbine companies' attempts to learn from the automotive industry have, until now, mainly been about how to apply cost-cutting. Getting to the level of sophistication of auto manufacturing, however, is about achieving a higher level of investment over a sustained period and not just about being leaner.

China versus the West – a market split in two?

One final issue that will have a lasting impact on wind-industry productivity and costs is the impact of the Chinese wind industry on the global market. The most striking feature of the current situation is that from one perspective, there are currently two almost separate wind markets: the Americas and Europe on the one hand and China on the other.

The European and American markets are dominated by the same group of international turbine manufacturers, led by Vestas, GE, Siemens, Suzlon–REpower, Enercon and Gamesa. They also share many of the same international wind developers, led by Iberdrola, EDPR, EDF-EN, E.ON and so on, although there are also many regionally focused ones too. The Chinese market is now almost completely dominated by Chinese turbine manufacturers, as we saw in Chapter 3, with the non-Chinese companies struggling to make it into the top ten spots. China's top ten wind developers are entirely Chinese, and none of the biggest international players have even attempted to break into the market. This is in contrast to the rest of the world, where a number of large-scale developers such as Iberdrola, EDPR, EDF-EN, Acciona, Enel and E.ON have projects across a wide range of countries.

The existence of two almost separate markets, particularly when China is consistently outstripping any other country or even region in terms of annual growth, has contributed to the growth of unsustainable levels of excess capacity. And it can only hold back achieving higher productivity and lower costs in the industry as a whole. Of course, the issue has more nuances than at first appear. Western companies that have lost market share or even given up on selling into the Chinese market at all often maintain large-scale manufacturing operations in the country.

One example is Nordex, which decided to pull back from the Chinese market after negotiations over a joint venture with a leading developer failed to prosper in 2012. The German company continues to use China as a major source for component supply, as do others which have pulled back from the market such as REpower.

The US's GE also maintains significant production activities in China, as well as maintaining a small share of the local market. An official from

GE commented recently in Beijing that 'You could be receiving GE turbines made in China and you wouldn't even know it' (personal communication). Other big producers, such as Gamesa, also source components from China for their global operations, particularly large, heavy parts such as hubs.

The divide between Chinese and non-Chinese OEMs is often not as wide as it may first appear, and questions of intellectual property rights are seldom as conflictive as they have been in the Sinovel–AMSC case. Goldwind, the Chinese company that is arguably best placed to expand internationally, bought into the highest level of turbine design through its acquisition of German designer and manufacturer Vensys, and licensees around the world are using designs whose patents are controlled by the company. Goldwind has also been the most successful of the Chinese companies in gaining experience in hard-to-enter 'mature' Chinese wind markets, and to all extents and purposes appears as a Western company in places like the US and Australia.

Meanwhile, Ming Yang is using designs from German company Aerodyn for its new turbines. Other companies have extensive collaboration with Western companies such as LM Windpower, The Switch or Mita-Technik on core parts of the turbine such as blades, drive trains and power electronics.

The push by Chinese turbine manufacturers to become international players will eventually materialise at some level, although analysts IHS-EER consider the first wave of 'go global' – in the 2011–13 period – 'to have failed'. Chinese utilities/developers such as Longyuan are slowly accumulating projects abroad and will give Chinese OEMs a foothold in these markets as they do so. Other important market participants such as State Grid are also stepping up their international presence. And the steady shift of financial power from the West to China – helped by developments such as the soon-to-be establishment of the renminbi as an international currency, and the availability of finance – will give Chinese companies a significant advantage in the years to come.

Meanwhile, the Western companies that have decided to stay and fight it out in China could eventually win back some market share, as developers focus on quality, and the costs of local producers rise, although opinion is divided on this, with 'don't hold your breath' the prevailing view from most analysts. GWEC's Steve Sawyer is sanguine about the prospects for Chinese companies abroad, arguing that expectations were much too high in the first place. 'The idea that Chinese turbine manufacturers were going to take over the world market was basically an expression of Sinophobia', he says. 'In the mid-term Chinese manufacturers will probably end up controlling about a similar proportion in international markets that non-Chinese manufacturers control in China.' (Comments made at breakfast that I hosted in October 2013 in Beijing. See http://www.rechargenews.com/wind/asia_australia/article1345485.ece.) This seems to be a sensible view, and at the present there is a long way to go before Chinese OEMs win market share of 10 per cent in major markets such as the US or Germany.

At present though, the divide between the Chinese and non-Chinese wind markets poses a number of problems for the wind industry. No major OEMs have been able to reap the potential scale advantage of rapid growth in all the major markets, and neither have we seen true global competition that could really raise productivity and lower costs. Instead, as we have seen, the Chinese market has until now spawned an artificially high number of new entrants, with many of these winning share despite low levels of technical expertise and quality. Non-Chinese companies have had to scale down their industrial presence in China, while Chinese players have, on the whole, struggled to gain credibility for their products abroad. This means that non-Chinese markets have not seen the potential cost benefits that Chinese industrial productivity and scale could bring (although it is a moot point if these benefits could be maintained if exporting to the West or producing outside China). Meanwhile China's market has yet to benefit from the experience and skills that developers used to making projects viable in highly competitive markets could bring.

Hopefully, the not too distant future will see a truly international turbine market, with highly capitalised and internationally diversified players that are able to compete for leadership in China and internationally. At the moment, this prospect still seems distant, but as the balance sheets of Chinese turbine companies return to health and executive management grows in sophistication, the prospect of major acquisitions or mergers that could create the next level of international manufacturer grows.

Notes

1 At the time this book was in production, however, GE did indeed seem poised to make a major acquisition of the power assets of Alstom. These include significant wind assets and could effectively propel GE back into the offshore wind space.

Conclusion
Who will reap the wind?

I would have to be a rather reckless observer to try to answer the question of what the future of the wind industry is and which countries and companies will dominate it.

As we have seen, the direction of our energy matrix remains the most political of questions and, ultimately, the prospects for how wind will develop are tied up with our expectations as to whether humanity is capable of solving some of the really big challenges of our time, including dangerous climate change and fossil-fuel dependency.

At the time of writing, the world seems headed for an unsustainable level of emissions, while politicians have so far proved incapable of creating stable international frameworks which would force a big enough change in investment patterns to reverse the trend. Indeed, recent years seem to have seen a collective loss of willpower around climate change and a resurgence in the most backward-looking trends in the energy industry. Europe, the global front-runner in fighting climate change and promoting renewables, is at a crossroads, as wind's emergence as a big-league power source has stirred up a hornet's nest of incumbent interests in the power industry and their political and bureaucratic allies.

On the other hand, wind has proved remarkably resilient, as can be seen for instance by its continuing growth in the US in the face of natural-gas prices dipping below US$2/MBTU, and has consistently outstripped the predictions of even its own industry bodies, not to mention those of institutions such as the IEA.

European power markets urgently need redesigning, but major governments such as that of Germany are unlikely to want to abandon long-held policy goals and establish a framework that would enshrine a comeback for imported fossil fuels.

Perhaps even more hopefully, the big developing countries such as China, India, Brazil and South Africa – the first two being climate negotiation 'villains' – along with countries like Saudi Arabia and Australia, are implementing policies that are ever more favourable to renewables, and huge new markets are being created. This will make retreating from climate policies even less tenable for the OECD countries, while creating growing opportunities for collaboration.

Despite the high level of political noise and hostile press, public support for wind remains strong across all major markets, and wind is likely to be able to create growing market leverage for its output as major corporations like Google, Facebook and Ikea continue to 'green' their operations, helped by industry initiatives like WindMade.

The arrival of some sort of 'grid parity' is not some kind of myth promoted by the renewables industry. In many parts of the world, wind and solar are already cheaper than fossil-fuel generation, even without properly factoring in the costs of carbon. Wind's ability to innovate, to overcome technical barriers and lower costs continues to be key, as it goes head-to-head with thermal and nuclear projects for a large piece of the global power industry.

How much of a piece is up for grabs? As well as competing on the market with both fossil and non-fossil generation sources, wind will need to continue to define its place in an emerging story that is likely to include the explosive growth of solar power, as well as the continuing existence of nuclear power and residual fossil-fuel production, hopefully using some degree of carbon sequestration.

How the wind-turbine industry evolves within the wider picture will depend primarily on the speed of growth. As a recent report from analysts MAKE Consulting pointed out, an uneven, stuttering level of growth is likely to lead to the survival of niche players concentrated in local or regional markets. Slower growth will also mean a continuing level of caution when it comes to investment, and hence a relatively slower growth in productivity.

A renewed spurt of sustained growth, however, will once again make the wind-turbine industry the subject of *Forbes* magazine's front pages and analyst 'buy' ratings. Visibility on growth along with increased cost and margin competition is likely to spur a long-awaited wave of merger and acquisitions activity and consolidation, and the emergence of a group of front runners, who are able to turn their global industrial footprints and strengths in knowhow and R&D into a price advantage. This process is likely to be given an added spur if and when the offshore sector, with its huge capital costs and risks, lives up to its growth expectations.

Whether the heavyweight group will be made up of four, six or eight companies is impossible to predict, and it would be unfair for me to lay bets on market winners. From today's vantage point, the big diversified industrial players – GE and Siemens – are likely be present in some form, along with industry pioneer Vestas, and at least two big Chinese players.

While they are minor players in the market today, it would be unwise to rule out Korea's Samsung and Japan's MHI as potential majors. There are also specialist players with major strengths in terms of technical knowhow, such as Spain's Gamesa and Germany's Enercon, as well as Suzlon–REpower.

Will EU governments and company officials eventually see the need for a 'European champion' along the lines of Airbus to face up to Asian competition? Are the medium-sized specialist companies in Europe willing to sell to bigger rivals and will we see a large-scale bankruptcy of a wind-turbine manufacturer? Will GE decide to make a serious push to build up its capacities outside the US through acquisition? Will Chinese or other Asian companies finally make the big European acquisitions that some have been predicting for years?

How the companies in this book will position themselves and relate to each other, how they succeed or fail in creating the products and technologies that can win market share and knock out competitors, will be a key part of the story of the rise of wind power in the coming months and years.

Postscript

At the time when *Wind Power* was being prepared for printing (July 2014), the trends that we addressed in the book seemed to be accelerating.

Wind-power generation continues to make steady inroads in the global power market. There has been a significant uptick in new projects in China and the US, while new markets in Latin America, Asia and Africa continue to pick up speed. Offshore, several landmark projects have been approved and are set to go ahead. Some of the wind-turbine manufacturers that were in serious trouble in 2012 and 2013 – particularly Vestas, the biggest company in the sector – continue to see their fortunes revive and are beginning to speed up their investments once again.

Widespread consolidation among wind-turbine manufacturers is finally occurring. The MHI–Vestas Offshore joint venture was formally launched on 1 April 2014 and seemed to be making steady progress in becoming a heavyweight competitor to Siemens. The company had been selected in February by DONG as its preferred supplier for its Burbo Bank Extension project and officials said that they expect several more orders to be announced soon.

Competitors Areva and Gamesa are, as we have seen, preparing to merge their offshore operations, with Areva strengthening its position through winning 1GW of orders in the second French offshore tender.

But the really big news has been the protracted battle between General Electric and Siemens to take over France's Alstom, which resulted in approval in June and success for GE. While wind has not been the main driver in the deal, the takeover is likely to have profound implications for the struggle to dominate wind.

At the time of writing, the takeover was set to see GE taking over Alstom's onshore wind operations, which would see it establish a commanding position in Brazil, and help it in its plans to become a serious player in Europe. The deal would also see GE and Alstom creating a joint venture in offshore wind, effectively putting GE back into the sector and creating another large and well-capitalised competitor to Siemens, to add to MHI–Vestas and Areva–Gamesa.

The debate about the business model for wind continues to evolve, as new models for financing and services emerge and the influence of disruptive technology change in other areas of economic life make themselves felt. Along with solar, the changing wind market is likely to play an ever greater role in the wider evolution of the global energy system in the coming months and years as the transition to a new, sustainable matrix continues.

References

Backwell, B. (2009a) Sector is hopeful, but warns of 'unravelling' if talks fail, *Recharge*, 19 November, 2009. Available online at: http://www.rechargenews. com/magazine/article1284642.ece.

Backwell, B. (2009b) UTC steps in to save Clipper from meltdown, *Recharge*, 17 December 2009. Available online at: http://www.rechargenews.com/magazine/ article1283348.ece.

Backwell, B. (2010a) Iberdrola chairman flies in as companies seek agreement, *Recharge*, 9 September 2010. Available online at: http://www.rechargenews.com/ magazine/article1288037.ece.

Backwell, B. (2010b) Tanti outlines strategy to become world number one, *Recharge*, 8 October 2010. Available online at: http://www.rechargenews.com/news/policy_ market/article1288358.ece.

Backwell, B. (2010c) Is Siemens line the future of turbine production?, *Recharge*, 23 July 2010. Available online at: http://www.rechargenews.com/wind/ article1286518.ece.

Backwell, B. (2011a) Indian court deals Enercon a blow over turbine patents, *Recharge*, 4 February 2011. Available online at: http://www.rechargenews.com/ news/policy_market/article1289702.ece.

Backwell, B. (2011b) Gamesa seals $2bn turbine coup with India's Caparo, *Recharge*, 17 May 2011. Available online at: http://www.rechargenews.com/wind/ article1291277.ece.

Backwell, B. (2011c) Indian wind market on course to break 3GW barrier this year, *Recharge*, 25 November 2011. Available online at: http://www.rechargenews.com/ wind/article1294887.ece.

Backwell, B. (2012a) Blood on the carpet as Vestas restructures, *Recharge*, 19 January 2012. Available online at: http://www.rechargenews.com/wind/article1295345. ece.

Backwell, B. (2012b) Vestas in disarray: deputy CEO resigns ahead of results, *Recharge*, 7 February 2012. Available online at: http://www.rechargenews.com/ wind/article1299018.ece.

Backwell, B. (2012c) Vestas says former CFO did India deals without board approval, *Recharge*, 2 October 2012. Available online at: http://www.rechargenews.com/ wind/article1298501.ece.

Backwell, B. (2013a) Interview: UB Reddy, India's Wippa, *Recharge*, 5 July 2013. Available online at: http://www.rechargenews.com/wind/asia_australia/ article1330668.ece.

Backwell, B. (2013b) Brazil: The soft superpower, *Recharge*, 2 September 2013. Available online at: http://www.rechargenews.com/ThoughtLeaders/article1335038.ece.

Backwell, B. (2013c) Alstom: Brazil turbine costs will fall, *Recharge*, 12 September 2013. Available online at: http://www.rechargenews.com/wind/americas/article1337072.ece.

Backwell, B. (2013d) Gamesa expects Brazil shake out, *Recharge*, 11 September 2013. Available online at: http://www.rechargenews.com/wind/americas/article1337040.ece.

Backwell, B. (2013e) Vestas-Mitsubishi in pole position, *Recharge*, 1 November 2013. Available online at: http://www.rechargenews.com/wind/offshore/article1341789.ece.

Backwell, B. (2013f) GE prefers growth to acquisition, *Recharge*, 5 February 2013. Available online at: http://www.rechargenews.com/wind/europe_africa/article1316145.ece.

Backwell, B. (2013g) Vestas fraud probe centres around 'ghost' Chinese turbine purchase, *Recharge*, 28 May 2013. Available online at: http://www.rechargenews.com/wind/article1327963.ece.

Backwell, B. (2013h) Vestas momentum grows – analysts, *Recharge*, 3 July 2013. Available online at: http://www.rechargenews.com/wind/europe_africa/article1331380.ece.

Backwell, B. (2013i) MAKE: Orders bode well for 2014, *Recharge*, 26 September 2013. Available online at: http://www.rechargenews.com/wind/article1338759.ece.

Backwell, B. (2013j) Suzlon loses India wind top spot, *Recharge*, 10 May 2013. Available online at: http://www.rechargenews.com/wind/asia_australia/article1326241.ece.

Backwell, B. and Publicover, B. (2013) The future for foreign firms in China, *Recharge*, 4 October 2013. Available online at: http://www.rechargenews.com/magazine/article1338688.ece.

De Clercq, G. (2013) European utilities CEOs urge end to renewables subsidies, *Reuters*, 11 October 2013. Available online at: http://uk.reuters.com/article/2013/10/11/utilities-renewables-ceos-idUKL6N0I106Y20131011.

Dehlsen, J. (2003) Interview US Department of Energy. Available online at: http://www.windpoweringamerica.gov/filter_detail.asp?itemid=683.

EWEA (2010) EWEA 2010 statistics: Offshore and eastern Europe the new growth drivers for wind power in Europe, 31 January 2011. Available online at: http://www.ewea.org/news/detail/2011/01/31/ewea-2010-statistics-offshore-and-eastern-europe-new-growth-drivers-for-wind-power-in-europe/.

EWEA (2013b) *EU wind power grows in 2012, but industry challenged in 2013*. Available online at: http://www.ewea.org/news/detail/2013/02/08/eu-wind-power-grows-in-2012-but-industry-challenged-in-2013/.

EWEA Offshore Conference (2013) *Insight at Thought Leaders VIP brunch*. Available online at: http://www.ewea.org/offshore2013/wp-content/uploads/EWEA-Offshore-2013-day-three.pdf.

GWEC (2009) Renewable energy financing in a year of economic turmoil, *Global Wind 2009 Report*. Available online at: http://gwec.net/wp-content/uploads/2012/06/GWEC_Global_Wind_2009_Report_LOWRES_15th.-Apr..pdf.

GWEC (2011) *Global Wind Report 2011*. Available online at: http://gwec.net/wp-content/uploads/2012/06/Annual_report_2011_lowres.pdf.

GWEC (2012) *Analysis of the regulatory framework for wind power generation in Brazil*. Available online at: http://gwec.net/wp-content/uploads/2012/06/Brazil_report_2011.pdf.

Hoel, A. (2004) A wind Bonus for Siemens, *Power Engineering International*, 1 November 2004. Available online at: http://www.powerengineeringint.com/articles/print/volume-12/issue-11/regulars/news-analysis/a-wind-bonus-for-siemens.html.

Kessler, R. (2012) New Clipper owner shrinks workforce, looks to O&M, *Recharge*, 21 August 2012. Available online at: http://www.rechargenews.com/wind/article1298182.ece.

Leal, M. (2013a) Tolmasquim: 'It's wind's moment', *Recharge*, 2 September 2013. Available online at: http://www.rechargenews.com/wind/americas/article1334827.ece.

Leal, M. (2013b) Renova and the gold mine, *Recharge*, 3 September 2013. Available online at: http://www.rechargenews.com/wind/americas/article1335050.ece.

Lee, A. (2013a) Euro CEOs blast support for RE, *Recharge*, 11 October 2013. Available online at: http://www.rechargenews.com/wind/article1340231.ece.

Lee, A. (2013b) Suzlon agrees debt restructuring, *Recharge*, 24 January 2013. Available online at: http://www.rechargenews.com/wind/asia_australia/article1315033.ece.

Lee, A. (2014a) Siemens plants UK 'game changer', *Recharge*, 25 March 2014. Available online at: http://www.rechargenews.com/wind/article1356182.ece.

Lee, A. (2014b) Vestas EBIT beats expectations, *Recharge*, 3 February 2014. Available online at: http://www.rechargenews.com/wind/europe_africa/article1351179.ece.

Lee, A. and Jensen, K. (2012) Vestas seeks investor for up to 20% stake says chairman, *Recharge*, 24 August 2012. Available online at: http://www.rechargenews.com/wind/article1298204.ece.

Lee, A. and Publicover, B. (2014) Sinovel faces China criminal probe, *Recharge*, 13 January 2014. Available online at: http://www.rechargenews.com/wind/asia_australia/article1348792.ece.

Li, J. *et al.* (2012) *China Wind Energy Outlook 2012*. Available online at: http://www.gwec.net/wp-content/uploads/2012/11/China-Outlook-2012-EN.pdf.

Magee, D. (2009) *Jeff Immelt and the New GE Way*, McGraw Hill.

Mexia, A. (2011) *EDP-CTG Strategic Partnership Establishment Friday, 23rd December*. Available online at: http://www.edp.pt/en/Investidores/EDP%20Ficheiros/Transcript%20EDP-CTG%20Partnership.pdf.

Milne, R. (2012) *Vestas investors seek wind of change*, ft.com, 2 September 2012. Available online at: http://www.ft.com/cms/s/0/f4622308-f4dc-11e1-b120-00144feabdc0.html#axzz31lzk2Hzg.

Milne, R. (2013) *Vestas eyes 'future without surprises'*, ft.com, 21 August 2013. Available online at: http://www.ft.com/cms/s/0/641a1190-0a78-11e3-9cec-00144feabdc0.html#axzz31lzk2Hzg.

Mumma, C. (2002) GE seeks refund from Enron Wind, *Bloomberg News*, 15 November 2002. Available online at: http://articles.latimes.com/2002/nov/15/business/fi-wind15.

Nicola, S. and Bauerova, L. (2014) Dirtiest coal's rebirth in Europe flattens medieval towns, *Bloomberg*, 6 January 2014. Available online at: http://www.bloomberg.com/news/2014-01-06/dirtiest-coal-s-rebirth-in-europe-flattens-medieval-towns.html.

Nordex (2012) *2012 Annual Report Nordex SE*. Available online at: http://www.nordex-online.com/fileadmin/MEDIA/Geschaeftsberichte/Nordex_GB2012_en.pdf.

Patton, D. (2012) Small producers put the heat on China's wind big-guns, *Recharge*, 8 June 2012. Available online at: http://www.rechargenews.com/wind/article1297677.ece.

Patton, D. (2013) The rise and stall of Sinovel, *Recharge*, 6 January 2013. Available online at: http://www.rechargenews.com/magazine/article1302268.ece.

Publicover, B. (2013a) Marubeni seeks early view, *Recharge*, 7 August 2013. Available online at: http://www.rechargenews.com/wind/asia_australia/article1333936.ece.

Publicover, B. (2013b) Sinovel Flags Offshore Fight Back, *Recharge*,17 October 2013. Available online at: http://www.rechargenews.com/wind/asia_australia/article1340641.ece.

Ramesh, M. (2012) Vestas to scale down India operations, *The Hindu*, 25 September 2012. Available online at: http://www.thehindubusinessline.com/companies/vestas-to-scale-down-india-operations/article3935285.ece.

Recharge (2013) The Rise and Stall of Sinovel, *Recharge*, 6 January 2013. Available online at: http://www.rechargenews.com/magazine/article1302268.ece.

Recharge (2013) Suzlon woes 'dragging on Repower', *Recharge*, 14 August 2013. Available online at: http://www.rechargenews.com/wind/europe_africa/article1334457.ece.

Recharge (2013) Vestas momentum grows – analysts, *Recharge*, 3 July 2013. Available online at: http://www.rechargenews.com/wind/europe_africa/article1331380.ece.

Recharge (2014) JV is a sign of the times, *Recharge*, 17 January, 2014. Available online at: http://www.rechargenews.com/wind/offshore/article1349410.ece.

Singh, C. (2013) HSBC, 24 June 2013.

Snieckus, D. (2013) Samsung's offshore giant, *Recharge*, 5 November 2013. Available online at: http://www.rechargenews.com/wind/offshore/article1341693.ece.

Stromsta, K.E. (2009) Suzlon asks the billion-dollar question, *Recharge*, 17 December 2009. Available online at: http://www.rechargenews.com/magazine/article1283355.ece.

Stromsta, K.E. (2012) In Depth: GE's European push, *Recharge*, 3 December 2012. Available online at: http://www.rechargenews.com/wind/europe_africa/article1345356.ece.

Stromsta, K.E. (2013a) Interview: Ignacio Martín, Gamesa, *Recharge*, 7 June 2013. Available online at: http://www.rechargenews.com/magazine/article1327930.ece.

Stromsta, K.E. (2013b) MRP boss hails Marubeni link, *Recharge*, 5 August 2013. Available online at: http://www.rechargenews.com/wind/article1333817.ece.

Stromsta, K.E. (2013c) Vestas initiates lawsuits in India, *Recharge*, 26 July 2013. Available online at: http://www.rechargenews.com/wind/europe_africa/article1333196.ece.

Tanti, T. (2012) Opinion: Suzlon Chairman Tulsi Tanti, *Recharge*, 6 January 2012. Available online at: http://www.rechargenews.com/magazine/article1302254.ece.

Vestas (2004) Press release 8 July 2004. Available online at: http://www.vestas.com/files/Filer/EN/Press_releases/VWS/2004/080704-UK.pdf.

Williams, M. and Menon, M. (2013) Bewildering Indian policies fuel needless coal imports, *Reuters*, 13 October 2013. Available online at: http://www.reuters.com/article/2013/10/13/us-india-coal-idUSBRE99C0CB20131013.

Windpower Monthly (2001) Two wind giants go head to head – Vestas and Gamesa split, 1 January 2002. Available online at: http://www.windpowermonthly.com/article/950913/two-wind-giants-go-head-head---vestas-gamesa-split.

Windpower Monthly (2004) Danish wind giants head for merger – Vestas and NEG Micon going for global supremacy. 1 January 2004. Available online at: http://www.windpowermonthly.com/article/956160/danish-wind-giants-head-merger---vestas-neg-micon-going-global-supremacy.

Yergin, D. (2011) *The Quest: Energy, Security and the Remaking of the Modern World*, Allen Lane.

Index

Page numbers for figures are shown in *italic*; tables are shown in **bold**.